Methods and Tools in Biosciences and Medicine

Microinjection,
edited by Juan Carlos Lacal, Rosario Perona and James Feramisco, 1999
DNA Profiling and DNA Fingerprinting,
edited by Jörg Epplen and Thomas Lubjuhn, 1999
Animal Toxins – Facts and Protocols,
edited by Hervé Rochat and Marie-France Martin-Eauclaire, 1999

Microinjection

Edited by

Juan Carlos Lacal
Rosario Perona
James Feramisco

Birkhäuser Verlag
Basel · Boston · Berlin

Editors

Dr. Juan Carlos Lacal
Inst. de Investigaciones Biomedicas
Calle Arturo Duperier, 4
E-28029 Madrid
Spain

Dr. Rosario Perona
Inst. de Investigaciones Biomedicas
Calle Arturo Duperier, 4
E-28029 Madrid
Spain

Dr. James Feramisco
University of California Cancer Center
La Jolla, CA 92093–0636
USA

Library of Congress Cataloging-in-Publication Data
Microinjection / edited by Carlos Juan Lacal, Rosario Perona, James Feramisco.
    p.   2cm. - - (Methods and tools in biosciences and medicine)
  Includes bibliographical references and index.
  ISBN 0-8176-5973-0 (hardcover : alk. paper). - - ISBN 3-7643-5973-0
(hardcover : alk. paper). - - ISBN 0-8176-6019-4 (wire-o binding :
alk. paper). - - ISBN 3-7643-6019-4 (wire-o binding : alk. paper)
    1. Microinjections. 2. Cytology- -Technique. I. Lacal, Juan
Carlos. II. Perona, Rosario, 1956–   . III. Series.
  [DNLM: 1. Cytological Techniques. 2. Microinjections- -methods.
3. Genetic Engineering- -methods. 4. Transformation, Genetic-
-genetics. QH 585 M6263 1999]
571.6'07'2- -dc 21
DNLM/DLC
for Library of Congress   98-43502
                                CIP

QH
585
.5
M52
M53
1999

Deutsche Bibliothek Cataloging-in-Publication Data
Microinjection / ed. by: Juan Carlos Lacal ... – Basel ; Boston ;
Berlin : Birkhäuser, 1999
  (Methods and tools in biosciences end medicine)
  ISBN 3-7643-6019-4 (Basel ...)
  ISBN 0-8176-6019-4 (Boston)
  ISBN 3-7643-5973-0 (Basel ...)
  ISBN 0-8176-5973-0 (Boston)

The publisher and editor can give no guarantee for the information on drug dosage and administration contained in this publication. The respective user must check its accuracy by consulting other sources of reference in each individual case.

The use of registered names, trademarks etc. in this publication, even if not identified as such, does not imply that they are exempt from the relevant protective laws and regulations or free for general use.

© 1999 Birkhäuser Verlag, PO Box 133, CH-4010 Basel, Switzerland
Printed on acid-free paper produced from chlorine-free pulp. TCF ∞
Printed in Germany
ISBN 3-7643-6019-4   ISBN 3-7643-5973-0
ISBN 0-8176-6019-4   ISBN 0-8176-5973-0
9 8 7 6 5 4 3 2 1

Contents

List of Contributors

SOHAIL AHMED, Department of Neurochemistry, Institute of Neurology, London, UK

LIZABETH A. ALLISON, College of William & Mary, Department of Biology, Williamsburg, VA, USA

ROSARIO ARMAS-PORTELA, Centro de Biologia Molecular „Severo Ochoa" CSIC-UAM, Facultad de Ciencias, Universidad Autonoma de Madrid, Madrid, Spain

JESUS AVILA, Centro de Biologia Molecular „Severo Ochoa" CSIC-UAM, Facultad de Ciencias, Universidad Autonoma de Madrid, Madrid, Spain

FRANCISCO BARROS, Departamento de Bioquimica y Biologia Molecular, Facultad de Medicina, Universidad de Oviedo, Oviedo, Spain

AMANCIO CARNERO, Institute of Child Health, Cancer Biology Unit, University College London, London, UK

JOACHIM CLEMENT, Abteilung Biochemie, Universität Ulm, Ulm, Germany

FABRIZIO DOLFI, Instituto de Investigaciones Biomédicas, CSIC, Madrid, Spain

DIRCK EICK, GSF-Forschungszentrum, Institut für Klinische Molekularbiologie und Tumorgenetik, München, Germany

JAMES FERAMISCO, Dpts. Medicine and Pharmacology, School of Medicine, UCSD Cancer Center, San Diego, La Jolla, CA, USA

HENNER FRIEDLE, Abteilung Biochemie, Universität Ulm, Ulm, Germany

RANDY GAUGLER, Department of Entomology, Cook College, Rutgers University, New Brunswick, NJ, USA

GWYN W. GOULD, Division of Biochemistry and Molecular Biology, Institue of Biomedical and Life Science, University of Glasgow, Glasgow, Scotland

GHAZALA HASHMI, Department of Entomology, Cook College, Rutgers University, New Brunswick, NJ, USA

SARWAR HASHMI, Department of Entomology, Cook College, Rutgers University, New Brunswick, NJ, USA

HEIKO HERMEKING, Johns Hopkins Oncology Center, Baltimore, Maryland, USA

HARLAM E. IVES, Departments of Medicine and Cardiovascular Research Institute, University of California, San Francisco, CA, USA

WALTER KNÖCHEL, Abteilung Biochemie, Universität Ulm, Ulm, Germany

ERIK S. KNUDSEN, Department of Cell Biology, Neurobiology and Anatomy, University of Cincinnati, College of Medicine, Cinncinati, OH, USA

FRANZ KOHLHUBER, GSF-Forschungszentrum, Institut für Klinische Molekularbiologie und Tumorgenetik, München, Germany

ROBERT KOZMA, Department of Neurochemistry, Institue of Neurology, London, UK

MICHAEL KÜHL, Abteilung Biochemie, Universität Ulm, Ulm, Germany

JUAN CARLOS LACAL, Instituto de Investigaciones Biomédicas, CSIC, Madrid, Spain

JOHN C. LAWRENCE, JR., School of Medicine, Dpts of Pharmacology and Medicine, University of Virginia, Charlottesville, VA, USA

THOMAS LEUNG, Glaxo-IMCB Group, Institute of Molecular and Cell Biology, Singapore

LOUIS LIM, Department of Neurochemistry, Institute of Neurology, London, UK and Glaxo-IMCB Group, Institute of Molecular and Cell Biology, Singapore

JILL MANCHESTER, Department of Molecular Biology and Pharmacology, School of Medicine, Washington University, St. Louis, MO, USA

EDWARD MANSER, Glaxo-IMCB Group, Institute of Molecular and Cell Biology, Singapore

ROBERT S. MATHIAS, Children`s Renal Center, Department of Pediatrics, University of California, San Francisco, CA, USA

MONICA S. MURAKAMI, NCI Frederick Cancer Research and Development Center, Frederick, MD, USA

PILAR DE LA PENA, Departamento de Bioquimica y Biologia Molecular, Facultad de Medicina, Universidad de Oviedo, Oviedo, Spain

ROSARIO PERONA, Instituto de Investigaciones Biomédicas, CSIC, Madrid, Spain

MICHAEL J. SEATTER, Division of Biochemistry and Molecular Biology, Institue of Biomedical and Life Science, University of Glasgow, Glasgow, Scotland

MICHAEL WALTER, Abteilung Biochemie, Universität Ulm, Ulm, Germany

PAUL A. WALTON, Dept. of Anatomy and Cell Biology, The University of Western Ontario, London, Canada

DORIS WEDLICH, Abteilung Biochemie, Universität Ulm, Ulm, Germany

GEORGE VANDE WOUDE, NCI Frederick Cancer Research and Development Center, Frederick, MD, USA

1 Needle Microinjection: A Brief History

J. Feramisco, R. Perona and J.C. Lacal

Since its inception in the early 1900's (Barber, 1911), the technique of needle microinjection has become a prominent experimental approach in biological research. Cellular organelles, DNA and RNA, enzymes, structural proteins, metabolites, ions and antibodies are just some of the molecular and cellular elements that have been transposed from test tubes into living cells by needle injection. A simple search of the literature indicates the growing popularity of the technique, returning a few citations in the 1970's, and thousands in the 1990's. As a complementary approach to DNA transfection, and as one of only a few viable ways to introduce non-genetic, large molecules into living cells, microinjection is now routinely used to study many living cell systems. With an inherent immediacy, microinjection has facilitated a wide range of biological studies, some of which are described in this book. While numerous alternative "microinjection" tools have also been developed, such as vesicle fusion, scrape loading and electroporation, we will focus mainly on the versatility of the use of needles to facilitate cellular injection. Lastly we apologize to all of our colleagues whose exciting work could not be mentioned, and for any inadvertent errors of fact we may have made in this brief chapter. Many additional references can be found in the ensuing chapters, however, and we look forward to new innovations in the field from the readers of this book.

A logical adaptation of the pioneering use of micropipettes in electrophysiology in earlier years, micro needle injection of such organisms as amebae (Flickinger, 1974; Hawkins, 1969; Jeon et al., 1967) and mouse embryos (Lin, 1966) was already a field of research in the 1960's. While these organisms facilitated key biological studies in this time period, in the ensuing 10-15 years, investigators began to explore the possibility of using microinjection to deliver organelles and molecules to other cells such as paramecium (Knowles, 1974; Koizumi, 1974), frog and mouse oocytes and eggs (Gurdon et al., 1971; Gurdon, 1971; Heidemann and Kirschner, 1975; Krieg et al., 1984; Wasserman and Masui, 1976), and mammalian cells of somatic origin (Graessmann et al., 1971). These early studies (Graessmann et al., 1971; Gurdon et al., 1971; Stacey and Allfrey, 1977; Diacumakos and Gershey, 1977; Graessmann and Graessmann, 1976; Soreq, 1985; Capecchi, 1980) set the stage for many eventual uses of microinjection in biological discovery, embryo injection for the production of "transgenic" animals (Gordon et al., 1980; Harbers et al., 1981; Constantini and Lacy, 1980; Lacy et al., 1983; Brinster et al., 1981; Palmiter et al., 1982; Voellmy and Rungger; 1982; Wagner et al., 1981a; Wagner et al., 1981b; Wasserman and Masui, 1976; Wilson et al., 1972).

Methods and Tools in Biosciences and Medicine
Microinjection, ed. by J. C. Lacal et al.
© 1999 Birkhäuser Verlag Basel/Switzerland

In early microinjection experiments with *Xenopus* oocytes, the technique was initially used to study gene expression relating to the molecular regulation pathways of oocyte maturation (Gurdon, 1971). Subsequently, the system was exploited for other purposes, including the identification of receptors and ion channels of cloned genes from mammalian cells. In fact oocytes as microinjection vehicles were instrumental in enabling expression cloning of numerous important neuronal receptors. By measuring newly expressed receptor activity following injection of large pools of cDNAs from mammalian cells, isolation or identification of desired clones was accomplished.

Gurdon and others began to use *Xenopus* oocytes as recipients for microinjected molecules (Gurdon, 1971), and developed assays to study the influence of gene expression on the maturation pathways important to these cells. It was soon recognized that oocytes would be useful in unraveling intracellular signaling pathways related to hormone action in the maturation process. Studies of protein kinases and inhibitors (Maller et al., 1978), maturation promoting factor (Wasserman and Masui, 1976), and meiotic machinery (Heidemann and Kirschner, 1975) exemplify this phase of work. By the 1980's, genes of distinct function isolated from *Xenopus* and other animal cells were studied by microinjection into oocytes (Scheer et al., 1984), including secretory products (Kreig et al., 1984), various membrane receptors (Barnard et al., 1987; Julius et al., 1988; Lubbert et al., 1987; Masu et al., 1987; Noda et al., 1986; Pritchett et al., 1988; Tempel et al., 1988) and heat shock proteins (Bienz and Pelham,1982). This was possible due to the transcriptional and translational fidelity displayed by these living eggs when injected with foreign DNA and RNA, and their large size which allowed recovery of newly synthesized gene products. Mammalian oncogenic pathways were also analyzed in this system, beginning with Ras (Birchmeier et al., 1985; Lacal et al., 1987) and Mos (Sagata et al., 1988) oncogenes, where their effects on oocyte maturation and phospholipid turnover were analyzed.

This system became a "gold standard" for identifying the function of mammalian genes encoding neurobiologically important molecules, such as Na^+ and K^+ channels and GABA transporters, for example, because of their lack of these membrane activities prior to injection of the foreign genes. Due in part to this fact, oocyte injection systems were helpful for the cloning of several of these key genes by screening cDNA libraries or by testing candidate cDNA clones for activity following expression in oocytes. Some examples of this are serotonin receptors (Julius et al., 1988; Lubbert, 1987; Pritchett et al., 1988), substance K receptors (Masu et al., 1987), and as Na^+ and K^+ channels (Noda et al., 1986; Stanker et al., 1989; Stuhmer, 1989).

In somatic cells, initial experiments in the 1970's were also related to gene expression. In work throughout the early part of this decade, Graessmann studied the transformation activity of elements within viral genomes (Graessmann and Graessmann, 1976) and Stacey and co-workers examined globin mRNA expression (Stacey and Allfrey, 1976). Stacey also injected radiolabeled proteins and followed their fate by autophagy (Stacey and Allfrey, 1977), a process of intracellular degradation, and Capechi (Capecchi, 1980) and Graessmann (Graessmann et al., 1979) created stable cell lines expressing foreign genes by this high efficiency approach. Brinster and colleagues injected mRNA into mouse ovum as well (Brinster et al., 1980). The fact that covalently modified proteins could indeed be injected into living cells, and their fate followed, gave rise to experiments to define the intracellular dynamics of discrete proteins in cells by using fluorescently labeled entities. Initial work with fluorescently labeled cytoskeletal proteins such as actin (Kreis et al., 1979; Wang and Taylor, 1979) and alpha-actinin (Feramisco, 1979) facilitated visualization of intracellular protein dynamics during spreading of fibroblasts and development in sea urchin eggs.

Experiments with the viral oncoprotein T-antigen (Tjian et al., 1978) showed that growth control of mammalian cells could be studied with microinjected proteins, and set the stage for a burst of studies on cell cycle regulation with cellular protooncogenes in later years. The latter experiments in the 1980's led to the demonstration of the oncogenic effect of mutant Ras (Feramisco et al., 1984; Stacey et al., 1984) and E1A oncogene product (Krippl et al., 1984), and the dual role of Ras in both differentiation and growth (Bar-Sagi and Feramisco, 1984). Injection of inhibitory antibodies to signaling proteins to assess function was also established as a viable technique in this time period, where Tag (Antman and Livingston, 1980; Mercer et al., 1983), Mos (Stanker et al., 1983) and Ras (Mulcahy et al., 1985; Feramisco et al., 1985) were initially targeted. In fact, one of the most widely quoted studies in the field of growth regulation came from this approach where the requirement for cellular Ras function in serum stimulation of growth was described (Mulcahy et al., 1985).

Into the late 1980's many groups began using microinjection into animal cells to assay intracellular signaling molecules for effects on specific targets (Pasti et al., 1986). In the case of oncogene proteins, effects on the expression of endogenous were pioneered, where cellular Fos was examined following injection of Ras (Stacey et al., 1987), and in the case of the cAMP-dependent protein kinase, effects on both Fos and chimeric promoter-reporter genes were studied (Riabowol et al., 1988). Mammalian cells and microinjection were used to define the biological function of actin (Scheer et al., 1984) and of Rac, Rho and CDC42 (Ridley and Hall,1992; Kozma et al., 1995) in membrane ruffling, actin stress fiber and filopodia formation, work which stimulated many investigators to employ this technique in studies of the functions of related small "G"-proteins in protein trafficking and membrane activation in various types of animal cells.

Today, microinjection is broadly used as a valuable tool for the study of many different cell responses in a variety of systems. From individual cell signaling responses such as regulation of metabolic pathways and second function of second messengers, the fate of injected proteins in cellular compartments and cytoskeletal regulation, to definition of apoptosis and survival signals and transformation of whole organisms such as nematodes. Only a few of these applications are reported in this book to provide a sense of its power.

What does the future hold for this still-developing technique? We can predict that microinjection will continue to provide an important means of introduction of purified molecules into living cells, and a way to explore resultant cellular phenotypes. While limited by the number of cells studied and the lack of renewal of the injected substance, the advent of automated injectors and micro analytical techniques such as PCR will greatly add to the scope of possible microinjection experiments when combined. We look forward to the day when microinjection becomes sufficiently automated so that the great task of assigning cellular functions to the 100000 or so genes coming from the Genome Project might be enhanced. We also envision the exciting prospect of combining "microchip" cDNA methodologies with microinjection to facilitate the direct identification of all genes regulated by individual signal transduction molecules, including those appropriately post-translationally modified prior to injection. The future for this technology looks bright indeed.

References

Antman KH, Livingston DM (1980) Intracellular neutralization of SV40 tumor antigens following microinjection of specific antibody. Cell 19: 627-635.

Barber MA (1911) A technique for the innoculation of bacteria and other substances into living cells. J. Infec. Diseases 8: 348-352.

Barnard EA, Bilbe G, Houamed K, Moss SJ, Van Renterghem C, Smart TG (1987) Functional expression in the Xenopus oocyte of messenger ribonucleic acids encoding brain neurotransmitter receptors: further characterisation of the implanted GABA receptor. Neuropharmacology 26: 837-844.

Bar-Sagi D, Feramisco JR (1985) Microinjection of the ras oncogene protein into PC12 cells induces morphological differentiation. Cell 42: 841-848.

Bienz M, Pelham HR (1982) Expression of a Drosophila heat-shock protein in Xenopus oocytes: conserved and divergent regulatory signals. Embo Journal 1: 1583-1588.

Birchmeier C, Broek D, Wigler M (1985) Ras proteins can induce meiosis in Xenopus oocytes. Cell 43: 615-621.

Brinster RL, Chen HY, Trumbauer ME, Avarbock MR (1980) Translation of globin messenger RNA by the mouse ovum. Nature 283: 499-501.

Brinster, R., Chen, H., Trumbauer, M., Senear, A., Warren, R., and Palmiter, R (1981) Somatic expression of herpes thymidine kinase in mice following injection of a fusion gene into eggs. Cell 27: 223-231.

Capecchi MR (1980) High efficiency transformation by direct microinjection of DNA into cultured mammalian cells. Cell 22: 479-488.

Constantini F, And Lacy E (1981) Introduction of a rabbit globin gene into the mouse germ line. Nature 294: 92-94.

Diacumakos EG, Gershey EL (1977) Uncoating and gene expression of simian virus 40 in CV-1 cell nuclei inoculated by microinjection. Journal of Virology 24(3): 903-906.

Feramisco JR (1979) Microinjection of fluorescently labeled alpha-actinin into living fibroblasts. Proc Natl Acad Sci USA 76: 3967-3971.

Feramisco JR, Clark R, Wong G, Arnheim N, Milley R, McCormick F (1985) Transient reversion of ras oncogene-induced cell transformation by antibodies specific for amino acid 12 of ras protein. Nature 314: 639-642.

Feramisco JR, Gross M, Kamata T, Rosenberg M, Sweet RW (1984) Microinjection of the oncogene form of the human H-ras (T-24) protein results in rapid proliferation of quiescent cells. Cell 38: 109-117.

Flickinger CJ (1974) Radioactive labeling of the Golgi apparatus by micro-injection of individual amebae. Experimental Cell Research 88: 415-418.

Gordon JW, Scangos GA, Plotkin DJ, Barbosa JA, Ruddle FH (1980) Genetic transformation of mouse embryos by microinjection of purified DNA. Proc Natl Acad Sci USA 77: 7380-7384.

Graessmann M, Graessman A (1976) "Early" simian-virus-40-specific RNA contains information for tumor antigen formation and chromatin replication. Proc Natl Acad Sci USA 73: 366-370.

Graessmann, A, Graessmann, M, Topp, W., Botchan, M (1979) Retransformation of a simian virus 40 revertant cell line, which is resistant to viral and DNA infections, by microinjection of viral DNA. J. Virol. 32: 989-994.

Graessmann A, Graessmann M (1971) The formation of melanin in muscle cells after direct transfer of RNA from Harding-Passeg melanoma cells. Hoppe-Seyler's Z. Physiol. Chem. 352: 527-532.

Gurdon JB, Lane CD, Woodland HR, Marbaix G (1971) Use of frog eggs and oocytes for the study of messenger RNA and its translation in living cells. Nature 233: 177-182.

Gurdon JB (1973-74) Gene expression in early animal development: the study of its control by the microinjection of amphibian eggs. Harvey Lectures: 49-9.

Harbers K, Jahner D, Jaenisch R (1981) Microinjection of cloned retroviral genomes into mouse zygotes: integration and expression in the animal. Nature 293: 540-542.

Hawkins SE (1969) Transmission of cytoplasmic determinants in amoebae by micro-injection of RNA-containing fractions. Nature, 224: 127-129.

Heidemann SR, Kirschner MW (1975) Aster formation in eggs of *Xenopus laevis*. Induction by isolated basal bodies. J. Cell Biol. 67: 105-117.

Jeon KW, Lorch IJ, Moran JF, Muggleton A, Danielli JF (1967) Cytoplasmic inheritance in amoebae: modification of response to antiserum by micro-injection of heterologous cytoplasmic homogenates. Experimental Cell Research 46: 615-619.

Julius D, MacDermott AB, Axel R, Jessell TM (1988) Molecular characterization of a functional cDNA encoding the serotonin 1c receptor. Science 241: 558-564.

Knowles JK (1974) An improved microinjection technique in Paramecium aurelia. Transfer of mitochondria conferring erythromycin-resistance. Experimental Cell Research 88: 79-87.

Koizumi S (1974) Microinjection and transfer of cytoplasm in Paramecium. Experiments on the transfer of kappa particles into cells at different stages. Experimental Cell Research 88: 74-78.

Kozma R, Ahmed S, Best A, Lim L (1995) The Ras-related protein Cdc42Hs and bradykinin promote formation of peripheral actin microspikes and filopodia in Swiss 3T3 fibroblasts. Mol. Cell. Biol. 15: 1942-1952.

Kreis TE, Winterhalter KH, Birchmeier W (1979) *In vivo* distribution and turnover of fluorescently labeled actin microinjected into human fibroblasts. Proc Natl Acad Sci USA 76: 3814-3818.

Krieg P, Strachan R, Wallis E, Tabe L, Colman A (1984) Efficient expression of cloned complementary DNAs for secretory proteins after injection into *Xenopus* oocytes. J. Molec. Biol. 180: 615-643.

Krippl B, Ferguson B, Rosenberg M, Westphal H (1984) Functions of purified E1A protein microinjected into mammalian cells. Proc Natl Acad Sci USA 81: 6988-6982.

Lacal JC, de la Pena P, Moscat, Garcia-Barreno P, Anderson PS, Aaronson SA (1987) Rapid stimulation of diacylglycerol production in *Xenopus* oocytes by microinjection of H-ras 21. Science 238: 533-536.

Lacy E, Roberts S, Evans EP, Burtenshaw MD, Costantini FD (1983) A foreign beta-globin gene in transgenic mice: integration at abnormal chromosomal positions and expression in inappropriate tissues. Cell 34: 343-348.

Lin TP (1966) Microinjection of mouse eggs. Science 151: 333-337.

Lubbert H, Hoffman BJ, Snutch TP, van Dyke T, Levine AJ, Hartig PR, Lester HA, Davidson N (1987) cDNA cloning of a serotonin 5-HT1C receptor by electro-physiological assays of mRNA-injected *Xenopus* oocytes. Proc Natl Acad Sci USA 84: 4332-4336.

Maller JL, Kemp BE, Krebs EG (1978) *In vivo* phosphorylation of a synthetic peptide substrate of cyclic AMP-dependent protein kinase. Proc Natl Acad Sci USA 75: 248-251.

Masu Y, Nakayama K, Tamaki H, Harada Y, Kuno M, Nakanishi S (1987) cDNA cloning of bovine substance-K receptor through oocyte expression system. Nature 329: 836-838.

Mercer WE, Nelson D, Hyland JK, Croce CM, Baserga R (1983) Inhibition of SV40-induced cellular DNA synthesis by microinjection of monoclonal antibodies. Virology 127: 149-158.

Mulcahy LS, Smith MR, Stacey DW (1985) Requirement for ras proto-oncogene function during serum-stimulated growth of NIH 3T3 cells. Nature 313: 241-243.

Noda M, Ikeda T, Suzuki H, Takeshima H, Takahashi T, Kuno M, Numa S (1986) Expression of functional sodium channels from cloned cDNA. Nature 322: 826-828.

Palmiter RD, Chen HY, Brinster RL (1982) Differential regulation of metallothionein-thymidine kinase fusion genes in transgenic mice and their offspring. Cell 29: 701-710.

Pasti G, Lacal JC, Warren BS, Aaronson SA, Blumberg PM (1986) Loss of mouse fibroblast cell response to phorbol esters restored by microinjected protein kinase C. Nature 324: 375-377.

Pritchett DB, Bach AW, Wozny M, Taleb O, Dal Toso R, Shih JC, Seeburg PH (1988) Structure and functional expression of cloned rat serotonin 5HT-2 receptor. Embo Journal 7: 4135-4140.

Riabowol K, Fink S, Gilman M, Walsh D, Goodman R, Feramisco JR (1988) The catalytic subunit of cAMP-dependent protein kinase induces expression of genes containing cAMP-responsive enhancer elements, Nature 336, 83-86.

Ridley A, Hall A (1992) The small GTP ase Rho regulated the assembly of focal adhesions and stress fibers in response to growth factors. Cell 70: 389-389.

Sagata N, Oskarsson M, Copeland T, Brumbaugh J, Vande Woude GF (1988) Function of c-mos proto-oncogene product in meiotic maturation in Xenopus oocytes. Nature 335: 519-525.

Scheer U, Hinssen H, Franke WW, Jockusch BM (1984) Microinjection of actin-binding proteins and actin antibodies demonstrates involvement of nuclear actin in transcription of lampbrush chromosomes. Cell 39(1): 111-112.

Soreq H (1985) The biosynthesis of biologically active proteins in mRNA-microinjected Xenopus oocytes. Crc Critical Reviews in Biochemistry 18: 199-238.

Stacey DW, Allfrey VG (1977) Evidence for the autophagy of microinjected proteins in HeLA cells. J. Cell Biol. 75: 807-817.

Stacey DW, Allfrey VG (1976) Microinjection studies of duck globin messenger RNA translation in human and avian cells. Cell 9: 725-732.

Stacey DW, Watson T, Kung HF, Curran T (1987) Microinjection of transforming ras protein induces c-fos expression. Molecular and Cellular Biology 7: 523-527.

Stacey DW, Kung HF (1984) Transformation of NIH 3T3 cells by microinjection of Ha-ras p21 protein. Nature 310: 508-511.

Stanker LH, Gallick GE, Kloetzer WS, Murphy EC Jr, Arlinghaus RB (1983) P85: a gag-mos polyprotein encoded by ts110 Moloney murine sarcoma virus. J. Virology 45: 1183-1189.

Stuhmer W, Ruppersberg JP, Schroter KH, Sakmann B, Stocker M, Giese KP, Perschke A, Baumann A, Pongs O (1989) Molecular basis of functional diversity of voltage-gated potassium channels in mammalian brain. Embo Journal 8: 3235-44.

Tempel BL, Jan YN, Jan LY (1988) Cloning of a probable potassium channel gene from mouse brain. Nature 332: 837-839.

Tjian R, Fey G, Graessmann A (1978) Biological activity of purified simian virus 40 T antigen proteins. Proc Natl Acad Sci USA 75: 1279-1283.

Voellmy R, Rungger D (1982) Transcription of a Drosophila heat shock gene is heat-induced in Xenopus oocytes. Proc Natl Acad Sci USA 79: 1776-1780.

Wagner EF, Stewart TA, Mintz B (1981) The human beta-globin gene and a functional viral thymidine kinase gene in developing mice. Proc Natl Acad Sci USA 78: 5016-5020.

Wagner TE, Hoppe PC, Jollick JD, Scholl DR, Hodinka RL, Gault JB (1981) Microinjection of a rabbit beta-globin gene into zygotes and its subsequent expression in adult mice and their offspring. Proc Natl Acad Sci USA 78: 6376-80.

Wang YL, Taylor DL (1979) Distribution of fluorescently labeled actin in living sea urchin eggs during early development. J. Cell Biol. 81: 672-679.

Wasserman WJ, Masui Y (1976) A cytoplasmic factor promoting oocyte maturation: its extraction and preliminary characterization. Science 191: 1266-1268.

Wilson IB, Bolton E, Cuttler RH (1972) Preimplantation differentiation in the mouse egg as revealed by microinjection of vital markers. J. Embryol. Exp. Morphol. 27: 467-469.

Microinjection of Macromolecules into Mammalian Cells in Culture

R. Perona, F. Dolfi, J. Feramisco and J.C. Lacal

Contents

Methods and Tools in Biosciences and Medicine
Microinjection, ed. by J. C. Lacal et al.
© 1999 Birkhäuser Verlag Basel/Switzerland

1   Introduction

Microinjection of macromolecules into living cells has proved to be a powerful technique for the functional analysis of gene products in living somatic cells. This approach is useful for the introduction of functional proteins, genes, inhibitors of enzyme activities and antibodies into living cells (Ansorge, 1982). Types of cells into which these molecules can be introduced include primary and established cell lines, fibroblasts and cells of various tissue types, including contractile cardiac cells. Some of the advantages of microinjection over other methods of introducing biological materials into cells (Celis, 1984; Graessmann et al., 1980; Pepperkok et al., 1988) are as follows: (1) Economy of the material; microinjection delivers small volumes (10^{-13}–10^{-15} l) directly into specific cells which means that the amount of material needed is much less than would be required for other approaches (e.g., scrape loading). (2) Damage to the cell appears to be much less than with other methods as evidence by stress responses, etc. (3) There is much less limitation on the nature of the material injected; for example, nucleic acids, proteins, large and small molecules can all be introduced with little alteration of the technique. (4) Temporal experiments can be performed where the response at a given time to a particular reagent may be monitored. Also, two or more different proteins may be introduced consequently in a defined order into the cell by re-injecting the same cell more than once. It is not easy to do this by other approaches such as transfection. (5) Molecules which had been altered *in vitro* may be injected in order to determine the effect of post-translational modifications. (6) Direct comparison of the effect of the injected material may be made by comparing injected and uninjected cells or cells injected with control material on the same dish.

2   Materials

Equipment

The workstation for microinjection is composed of a microscope, an injecting pressure generator, and a micromanipulator. The system must be located in a room with low traffic and low wind to avoid cell culture contamination by microrganisms, as the microinjection experiments are carried out in open air.

Microscope

A good phase-contrast inverted microscope is a fundamental tool for successful microinjection experiments. The microscope must contain a manually or automated moving stage (depending upon the apparatus) with appropriately sized buckets to restrain the cell plates. Two optical systems are useful, one giving 200–320× magnification and one giving 100×. Microinjection is best done at 200 or 320× magnification, while the lower power is useful to drive the needle into the zone of cells to be injected. Furthermore, the optical set to work with fluorescent labels comprises an appropriate lamp and filter set depending on the fluorescent dye used, and may be on the same microcope or a different one. Semiautomatic and automatic systems are equipped with a video camera to display the cells under observation on a TV monitor. A balanced, vibration-free table is a fundamental component of microinjection systems as needle breakage and cell damage can be caused by vibrations.

Injecting pressure generator
A syringe connected by a Teflon tube to the pipet holder is the simplest positive injection pressure generator. However this system is not as easy to use as the automated pressure generators commercially available. Each of these systems can be coupled with several types of micromanipulators.

Micromanipulator
A wide range of micromanipulators is commercially available. Depending upon the study that will be carried out, choose the one that best fits your needs and budget.

Manual system: This is the simplest system, and is very useful and relatively inexpensive. This system allows a fine control of single cell injection; however, the number of injected cells in each experiment cannot be as high as with the use of an automated system. The simplest system is composed of a simple syringe as the pressure generator, connected to the pipet holder by a flexible Teflon tube. In this way the operator determines the positive pressure with his own hand. The pipet holder is attached to a manual micromanipulator, which is mounted next to an inverted microscope. This was the first system used in microinjection experiments. However the use of automatic pressure generators, commercially available, replaces some of the tedious and strenuous work for the operator.

Semiautomatic systems: This system still allows the operator to control single cell injections, but is faster than the manual one. The micromanipulator and the pressure source are connected and automated. The operator must first manually drive the needle to an area of cells to be injected. Next, the operator must set the needle coordinates (x, y and z), the pressure values and the injection time on a single cell and store the information into the system. These empirically determined values will be repeated on every cell the operator wants to inject by a simple press of a button.The semiautomatic system allows a high degree of reproducibility when the population of cells to be injected shows constant morphological characteristics, and the substrate on which the cells are attached is flat. These systems usually can also be used for manual injection, and they are more expensive than the manual ones.

Automatic systems: The automatic systems allow the operator to microinject more than a thousand cells in a short period of time, but the system is more complex and somewhat more difficult to control compared to the manual or semiautomatic systems. These systems employ computer-controlled positioning of the cells and the capillary and offer a good degree of reproducibility of the injection. Their cost is relatively high, however.

Solutions

Buffers for injection
Several kinds of buffers can be used in microinjection experiments. The injected solution must be both tolerated by the injected cell and compatible with the injected molecules. In general, detergents must be avoided, although we have successfully used small amounts of some weak detergents in the past. Moreover, it is very important to carefully control the amount of sample injected into the cell: Injection of more than 10% of the cell volume can cause cell death. Here we present some types of microinjection buffers (Table 1) that could be successfully introduced into most types of living cells without noticeable damage. Buffer composition may need to be changed depending on the nature of the protein to be injected, and each operator should experiment with his own formulation. Keep in mind, however, that appropriate controls for any buffer must be done along with experiments to ensure a lack of effects due to the buffer alone. Particular attention must be paid to preserve the activity of microinjected enzymes, and these activities should be re-checked following the employment of a microinjection buffer.

Table 1 Injection buffers

Sample	Buffer	Concentration
Proteins	(a) 20 mM MES pH 7.4 (b) 10 mM Tris-acetate pH 7.2 10 mM Sodium chloride or Potassium chloride 0.1 mM EDTA 0.1 mM 2-Mercaptoethanol 5–10% Glycerol.	1–10 mg/ml
Antibodies	10 mM Sodium phosphate pH 7.2 70–90 mM Potassium chloride	2 -10 mg/ml
DNA	10 mM Sodium phosphate pH 7.2 70–90 mM Potassium chloride	1–100 µg/ml

All the buffers must be prepared with bi-distilled water and sterilized by filtration (filter size: 0.2 µm).

- **Fixation Solution (Solution 1)**
    100 mM potassium phosphate, pH 7,4
    0,2% glutaraldehyde
    5 mM EGTA
    2 mM $MgCl_2$

- **Washing Buffer (Solution 2)**
  - 100 mM potassium phosphate, pH 7,4
  - 0,01% Na^+ desoxycholate
  - 0,2% Nonidet P40
  - 5 mM EGTA and 2 mM $MgCl_2$
- **β-gal Staining Buffer (Solution 3)**
  - 0.5 mg/ml X-gal, freshly prepared in dimethylsulfoxide or dimethylfor-
    mamide
  - 10 mM $K_3[Fe(CN)_6]$
  - 10 mM $K_4[Fe(CN)_6]$
  - all dissolved in **Solution 2**

3   Methods

3.1   General methods

Preparations of needle
In order to carry out successful microinjection experiments good capillaries are
essential. The needles must have an appropriate tip diameter, between 0.5 and
0.1 μm, to minimize both cell damage and clogging problems. Pre-pulled needles are commercially available, however several types of pullers, which are
also commercially available, allow one to easily obtain capillaries of appropriate
size. To obtain good microinjection needles, we suggest using borosilicate glass
thin wall capillaries (1.2 mm O.D., 0.69 I.D.). Available pullers must always be
standardized with respect to the type of glass used, tip diameter and shape.

Checking the needle
Each puller requires an appropriate setting that will change depending upon
the apparatus features, the type of glass used and the conditions (e.g. temperature, breezes) of the room in which the puller will be used. However, all the
pullers with which we have experience allow one to obtain good quality needles with comparable physical features.
    To check the quality of pulled capillaries a quick method is available.

Protocol 1   Checking the quality of pulled capillaries

1. Connect the pulled needle with the pressure source and apply a constant
   positive pressure.

2. Dip the needle in a transparent tube filled with ethanol.

   Note:
   If the tip is open you will be able to see a thin line of tiny bubbles emerging
   from the opening. If the tip diameter is too large, the bubbles will be very
   large and more evident. With practice the operator can learn to recognize
   the difference between a "good" needle and a "bad" one.

Opening the tip

Sometimes, depending upon the puller set up and the type of glass used, the needles are closed after the pulling procedure. Here we describe a simple method to open the tip.

Protocol 2   Opening the tip

1. Connect the needle with a pressure source and apply a constant positive pressure

2. Dip the needle tip for a short time (2–3 s) in a transparent tube containing 10% hydrofluoric acid (HF).

3. Dip the needle in a transparent tube containing ethanol to check if the needle is open.

4. If the needle is closed, dip the capillary first in distilled water and again in 10% HF and repeat as necessary.

Pipet filling

Two methods are available to fill the pipet: front filling and back filling. From our experience back filling is easier and faster than front filling, although front filling requires less material and thus also has advantages. In either case, we suggest that the sample be centrifuged for 10 min at 10000 rpm in microfuge just before beginning the microinjection experiments to eliminate large aggregates of sample which will lead to needle clogging. After ending the centrifugation, transfer some microliters of the sample to another tube. Be careful to minimize the movement of the tube and to aspirate only the upper part of the sample to avoid the small debris at the bottom of the centrifuge tube, which can lead to pipet clogging.

Protocol 3   Back filling of pipet

Glass capillaries with inner filament allow for easy backfilling. Typically about 0.1–1 µl of sample volume can be loaded into the rear opening of the needle. Sterile disposable microloaders are commercially available which have an extended tip to facilitate this process.

1. To load the needle, hold it in a vertical position (with the tip down) and insert the loader into the rear hole of the capillary.

2. Low the loader tip until it reaches the tip (constriction) in the pipet

3. Gently release the sample (to minimize air bubble formation) on the capillary wall and simultaneously remove the microloader from the glass capillary.

Note:

Often some air bubbles remain in the capillary tip. It is very important to re-move the bubbles because these will also lead to capillary clogging. To re-move the bubbles, keep the pipet in a vertical position with the tip down. Gently tap on the upper end of the glass capillary, which for capillaries with inner filaments will remove the bubbles.

Protocol 4   Front filling of pipet

Needles previously placed on the capillary holder of the micromanipulator can be loaded from the tip by aspirating the sample from a little liquid drop pre-viously placed on a non-wettable surface of a plate positioned under the micro-scope. This filling method avoids the problems of bubble formation and offers the opportunity to check the capillary tip opening directly under the micro-scope. On the other hand there can be the problem of capillary breakage con-sidering the small size of the drop of sample into which the needle must be placed for filling; indeed, it is very important to always know precisely where the plate surface is relative to the needle, considering that the needle tip is nor-mally damaged when it touches the plastic surface of the dish. Moreover, the filling operation must be fast in order to minimize evaporation of the sample since only small amounts of samples are usually available for placing onto the loading surface.

1. Place a clean tissue culture plate on the microscope stage and put the nee-dle into the pipet holder.

2. Put a small drop of the sample on the plate surface and focus the border of the drop into the magnified field of view. When the border is in focus, it will appear as a line; this border allows the operator to localize the plate sur-face.

3. Maintain this focus position and move the plate toward the center of the drop until a darker zone appears. This zone is the center of the drop, where the liquid thickness is greatest. It represents the zone for which the needlle is to be placed for loading.

4. Move the needle toward the sample on the plate, positioning its tip into this dark zone.

5. Lower the needle into the sample. When the capillary tip reaches the drop surface, the brightness of the liquid will change under the microscope.

6. Slowly lower the pipet and submerge it some micrometers into the drop. Be careful of keep the tip out of the focus (i.e., above the surface which is in fo-cus) to avoid touching the plate surface.

7. Without moving the pipet you can bring the tip of the needle into focus and fill the needle by applying a negative pressure.

8. The needle is ready for inmediate use.

3.2 Centering the capillary in the microinjection field

We briefly explained the major features of different microinjection systems: manual, semiautomatic and automatic systems. Considering the great differences between these, it is impossible to draw a general method to be used in microinjection experiments. The method will change depending upon the choice of apparatus used; however, we will attempt to show a general approach to be followed for any type of microinjection experiment.

It is important to consider that microinjection experiments are carried out in the open air, where the environmental conditions are different from the optimal ones of the incubator. To minimize cell stress and to carry out reproducible experiments it is important to keep the cell out of the incubator for a period of time not longer than 15 min. If longer periods of time are needed, replacing the medium on the cells with PBS, and using a 37°C stage warmer, can be helpful. In any case, it is important that the operator acquire a good familiarity with the apparatus to execute all the operations in as little time as possible before carrying out any microinjection experiments.

The operations to be followed in microinjection experiments are the following:
- Filling the capillaries (front or back filling)
- Bringing the needle into microscope view and centering the needle in the microinjection field
- (For semiautomatic and automatic apparatus) setup of device injection and movement parameters
- Injections

The first operation which is very important is to became familiar with the system. In order to rapidly carry out the experiments, practice centering the capillary tip in the microinjection field. This routine operation, which will become simple after practice, constitutes a frustrating obstacle to people with little or no experience. An incorrect operation or movement could easily lead to needle breakage and cell damage simply by touching the plate surface. Likewise, a delay in centering the needle and beginning the injections may result in needle clogging by aggregation of the sample.

Here, we present a simple method to carry out this operation minimizing the risks of capillary breakage. The method consists of different steps by which the operator can carefully lower the needle while maintaining its tip in the center of the microscope field. This method could be used successfully in manual, semiautomatic and automatic systems.

Protocol 5 Centering the capillary in the microinjection field

1. Focus the microscope on the cell surface using the working magnification (320×) and identify the grid on the substrate in the center of the slide. Choose the injection field in the cell monolayer.

2. Change to the lower power (100×) magnification and bring the cells into focus without moving the plate.

3. Move the capillary tip to the center of the field, but above the medium with the micromanipulator (the tip of the needle will appear as a bright unfocused object in the microscope).

4. Lower the needle carefully; notice that when the tip touches the liquid surface, the light brightness will change. It is important to remember that a low but constant pressure (holding pressure) should be applied to the needle so that medium will not flow into the sample.

5. As the microscope is still focused on the cell monolayer, the tip will appear as a shadow. It is crucial to get the tip of the needle into the field of view before lowering the needle any further. To identify the needle tip move the micromanipulator along the X and Y axis.

6. When the tip is identified, put it in the center of the field and change to 200× magnification.

7. Re-focus onto the cell monolayer and make sure that the needle tip is still centered.

8. Continue lowering the needle, but not quite all the way to its focus.

9. Change to the working magnification (320×) and re-focus on the cells. The needle tip will not yet be in focus but it is important to verify its position once again as above.

3.3 Cell microinjection

The microinjection method will change depending on the features of each apparatus. Here we describe a simple method to microinject using manual systems. We also briefly described the general issues to use semiautomatic and automatic systems. There are no technical differences between injection into cytoplasm and nucleus, except that for nuclear injections the amount of sample injected is less (about 1/10 of the liquid injected into cytoplasm).

Microinjection with manual systems
The manual microinjection systems is usually composed of an inverted phase contrast microscope, an injecting pressure generator and a micromanipulator.

The general features of the microscope have been described above. Several types of micromanipulators and injecting pressure generators are offered in an ever-growing market. For manual apparatus we suggest the use of a "joystick" micromanipulator. For the pressure generator it is best to use an apparatus which offers the following pressure ranges:

- **Holding pressure.** This provides a low pressure which prevents cell material or medium from entering the capillary tip. It must be regulated at low levels.
- **Injection pressure.** This provides a higher pressure required for injection.
- **Cleaning pressure.** This provides a brief but high pressure to be used for unplugging the capillary when it is clogged.
- **Vacuum pressure.** Negative pressure values to be used for the front filling of the capillary.

Protocol 6   Cell microinjection

1. Submerge the capillary in the cell medium; apply a cleaning pressure to begin the flow of sample through the narrow tip of the needle (focusing on the tip of the needle, the sample flow can often be seen as a phase dark stream flowing into the medium).

2. Re-focus microscope on the cell monolayer.

3. Set the holding pressure to allow for a low flow through the tip. The holding pressure prevents the medium from re-entering the capillary.

4. Set the injecting pressure and time by slowly lowering the needle into a "test" cell. For cytoplasmic injections, aim the needle at the thickest part of the cell, the area just next to the nucleus. Increase the pressure slightly to the "injection" pressure once inside the cell.

   Note:
   It is impossible to provide here any reference values as they will change depending on both the apparatus and the capillary features; however, injection times of less than 1 s are optimal to reduce cell damage and should easily be obtained if the flow of sample is sufficient. It is useful to know that when a cell is microinjected, tiny and gentle waves of phase dense fluid can be seen leaving the tip of the needle and flowing into the body of the cell or nucleus. If the flow is too great, the cell may explode; if the flow is too low, no injection will occur or the time necessary to transfer sample into the cell will be too great.
   Microinjection does cause some stress for the cell. Indeed, after injection cell refringence changes. If the injection caused excessive cell damage, within a few minutes of injection cell death (lysis) will be clearly evident.

5. When satisfactory settings are obtained, repeat as in 4., injecting the cells of the chosen field. With practice 100 or so cells can be injected within 15 min.

Alternatively, it is possible to carry out successful microinjection experiments by setting a holding pressure sufficient to maintain a constant flow through the needle tip. To inject a cell, touch it for a short period of time sufficient to transfer the appropriate amount of sample into the cell. Like the other method described it is possible to see a little wave of fluid entering the cell when the needle touches the cell. It is not necessary to change the microinjection pressure once it is calibrated on a few test cells as the liquid enters the cell as soon as the needle touches it. The amount of sample into the cell is determined mainly by the amount of time the needle tip remains in the cell.

Microinjection with semiautomatic apparatus
The fundamental basis of microinjection with semiautomatic systems is very similar to that described for manual systems. It varies only with different instruments from various manufacturers. For each apparatus the first step consists of setting up the conditions to microinject a single cell. This includes the following steps: 1) positioning of the needle over the cell, 2) setting the microinjection pressure and length of time for each injection, and 3) storing this information in the injection system. Once the conditions have been established, to carry out subsequent injections the operator simply guides the needle tip above a target cell in the field chosen for injection, and presses the activator switch. The switch or button then causes the instrument to inject the target cell. This method allows for injections of many cells in a few minutes.

Microinjection with automatic systems
As described in a previous chapter these systems allow the operator to inject a large number of cells in a short period of time. The system is controlled by a computer, and is operated with help of a video camera and monitor connected to the microscope. Machines vary depending upon the apparatus features purchased and the software used to control the system. As is the case for the semi-automatic system, the first step of operation consists of setting up the conditions of injection (position of the needle with respect to the cell, pressure and length of time of the injection). After storing the microinjection parameters the operator chooses by video monitor the cells and the intracellular zone to be injected, and the instrument will then carry out the injections. For both the automatic and semi-automatic injection systems, absolutely flat cell culture substrates are important so that the distance from the needle and the cells remains constant as the operator moves from field to field.

3.4   Needle clogging

One of the main problems connected with cell microinjection is needle clogging. In a previous paragraph we described how to minimize the clogging problems associated with capillary filling. Unfortunately, many times needle clogging arises during an experiment. Clogged needles have reduced or no sample flow through the tip, so that either the reproducibility of each injection is lost or no further injections can occur. This may require the operator to then increase the injection pressure during the experiment, or abandon it altogether to install another needle.

Here we provide some suggestions to help resolve this problem (when possible!).

- Sometimes it is possible to clean the needle by applying one or more times the cleaning pressure.
- If the needle remains clogged after above, move the needle to a field without cells. Lower the needle to allow it to gently touch the glass substrate and move it along a line perpendicular to the direction of the needle. It is very easy to carry out this operation if the cells are grown on glass coverslips. On the contrary it is difficult to move the needle on a plastic surface without breaking the tip.

Change the injection capillary if this operation also fails to clear the tip. If a cell sticks to the tip of the needle after injection, you may be able to remove it by rapidly raising the needle completely out of the medium, and then putting it back in.

3.5 Analysis of injected cells

Detection of injected products

Indirect immunofluorescence can be used to identify injected materials in different ways: identification of injected cells, determination of the amount of protein injected or expressed by the DNA injected and the determination of the intracellular localization of the injected gene product.

Protocol 7 Immunofluorescence staining of injected cells

1. Cells which have been plated on coverslips and injected are washed twice with phosphate-buffered saline (PBS)

2. Fix cells in 3,7% ($^V/_V$) formaldehyde in PBS for 30 min

3. Permeabilize cells with 0,1% ($^V/_V$) Triton X-100 for 5 min at room temperature (RT)

4. The coverslips are incubated in the presence of the primary antibody in PBS for 1 h at room temperature

5. Three washes of 15 min in PBS

6. The cells are then incubated with the secondary fluorochrome-conjugated antibody for 1 h at RT

7. Wash twice in PBS for 15 min

8. The coverslips are then mounted on glass microscope slides in a suitable mounting solution containing anti-photobleaching compounds.

The use of epitope tagged proteins is especially useful and effective in microinjection studies because it provides a direct way to distinguish between the injected and the endogenous protein. The epitope tagged proteins can be visualized using antibodies that recognize the added epitope. Among the more commonly used are hemagglutinin (HA) epitope that can be detected with monoclonal antibody 12CA5 (Boehringer Mannheim). The Myc epitope can be detected with the monoclonal antibody 9E10 (Santa Cruz Biotechnology). When expression plasmids are injected a sufficient level of expression can be detected at 2 h after injection.

Reporter assays: Gene transfer by microinjection is used not only to monitor the successful transfer of desired genes but also the establishment of the proper and predictable pattern of transgene expression. Some genes are well validated as reporter systems, such as ß-galactosidase, secreted alkaline phosphatase, chloramphenicol acetyltransferase and ß-glucuronidase (Riabowol et al., 1988; Alam and Cook, 1990; Martin et al., 1996, Bronstein et al., 1994). For these reporter genes detection of the activity of the promoter linked to the gene is usually quantified by the use of specific antibodies to the reporter (i.e. immunofluorescence) or in the case of ß-galactosidase, enzymatic activity. For ß-galactosidase enzymatic activity, advantage is taken of the stability of the bacterial enzyme to fixatives following expression in eukaryotic cells.

Protocol 8   β -galactosidase staining

1. Cells are washed in PBS (three times, taking great care if the cells are poorly attached to the substrate)

2. Fixation in **Solution 1** for 5 min at RT

3. Cells are washed three times with **Solution 2**

4. Solution 2 is replaced with **Solution 3**. Add sufficient solution 3 so the cells are covered and wait until the blue color appears.

Since color development is an enzymatic reaction, by measuring the kinetics of the appearance of blue color, estimates of the relative level of expression of reporter in cells can be made. This analysis should only be done to compare cells injected and stained in the same experiment, as color development will vary from experiment to experiment. Samples can later be stored at 4°C in 50% glycerol.

More recently the use of reporters based on firefly luciferase and green fluorescent protein (GFP), which allow the transgene expression to be very sensitively and non invasively measured, is greatly facilitating gene transfer technology. GFP of Aequorea is a 238 aminoacid-long polypeptide, and is highly fluorescent and stable in many assay conditions (Cubitt et al., 1995). GFP has excitation peaks at 395 nm (largest peak) and 474 nm and an emission peak at 509 nm (Prasher, 1995). GFP has been used in the measurement of gene expression, cell labeling and protein labeling localization studies. GFP shows no apparent toxicity, no apparent interference with normal cellular activities and is easy to assay using a fluorescence microscope. Modifications of GFP have been made using various mutagenesis schemes that improve fluorescence intensity, thermostability folding and formation of the chromophore (Crameri et al., 1996; Heim and Tsien, 1996). The ability to combine these modifications in synthetic GFPs has led to many additive gains. The S65T mutant is brighter and more resistant to photobleaching than wild type GFP (Heim and Tsien, 1996). Spectral variants (e.g. different colored GFPs) permit the simultaneous detection of expression from multiple reporters, tracking the transport and localization of more than one protein.

Cell proliferation
The synthesis of DNA in injected cells can be monitored by immunostaining of BrdU (bromodeoxyuridine, an analog of thymidine) incorporated into newly synthesized DNA.

Protocol 9   Immunostaining of BrdU

1. Following injection, BRdU is added to the medium at 10 mM and the cells incubated for 20 to 30 h

2. Fixing in 3,7% paraformaldehyde in PBS for 15 min

3. Treatment with 5 mM glycine and permeabilisation of the cells with 0.1% (V/V) Triton X-100 for 15 min

4. Coverslips are then incubated with the mouse monoclonal antibody specific for BrdU

Fluorescent labeled anti-mouse antibodies allow the identification of DNA synthesizing cells. This protocol can in principle be used for double staining of BrdU and the injected protein or gene product. Alternatively, injected cells can be incubated with tritiated thymidine for 20–30 h following injection to label newly synthesized DNA with this radioactive nucleotide. The incorporated radiolabel is then detected by emulsion autoradiography. In either case cells have to be able to utilize free nucleotides for DNA synthesis (e.g., thymidine kinase positive). While most studies of this type are aimed at determining whether or not cells have progressed through S-phase of the cell cycle, keep in mind that cells may incorporate the nucleotide labels into nuclear DNA for other reasons, such as by DNA repair.

Apoptosis assays
DNA fragmentation during apoptosis may lead to an altered nuclear morphology that can be easily identified under the microscope. A more reliable assay

for apoptosis is the enzymatic quantitation of the DNA breaks using the terminal deoxynucleotide transferase (TdT). The end labeling method known as TUNEL (TdT-mediated X-dUTP nick and labeling) has been modified to be used as an *in situ* assay by labeling the DNA with biotin-dUTP or DIG-dUTP, and then detecting the incorporated nucleotides in a second incubation step with streptavidin or an anti-DIG antibody conjugated with a reporter molecule (e.g. fluorescein, rhodamine, etc). Commercial kits of different suppliers are available for this assay.

References

Alan J and Cook JL (1990) Reporter genes: applications to the study of mammalian gene transcription. *Anal Biochem.* 188: 245–254.

Ansorge W (1982) Improved system for capillary microinjection into living cells. *Exp. Cell. Res.* 140, 31–37.

Bronstein I, Fortin J, Stanley PE, Stewart GSAB, Kricka LJ (1994) Chemiluminiscent and bioluminiscent reporter gene assays. *Anal Biochem* 219: 169–181.

Celis JE (1984) Microinjection of somatic cells with micropipets: comparison with other transfer techniques. *Biochem. J.* 223: 281–291.

Crameri A, Whitehorn EA, Tate E, Stemmer WPC (1996) Improved green fluorescent protein by molecular evolution using DNA shuffling. *Biotechnology* 14: 315–319.

Cubitt AB, Heim R, Adams SR, Boyd AE, Gross LA, Tsien RY (1996) Understanding improving and using green fluorescent proteins. *Trends Biochem. Sci.* 20: 448–455.

Graessmann A, Graessmann M, Mueller C (1980) Microinjection of early SV40 DNA fragments and T antigen. *Methods Enzymol.* 65: 816–825.

Heim R, and Tsien RY (1996) Engineering green fluorescent protein for improved brightness, longer wavelengths and fluorescence resonance energy transfer. *Curr Biol.* 6: 178–182.

Martin CS, Woght PA, Dobretsova A, Brotstein I (1996) Dual luminiscent-based reporter gene assay for luciferase and β-galactosidase. *Biotechniques* 21: 520–524.

Pepperkok R, Schneider C, Philipson L, Ansorge W (1988) Single cell assay with an automated capillary microinjection system. *Exp. Cell. Res.* 178, 369–376.

Prasher DC (1995) Using GFP to see the light. *Trends Genet.* 11: 320–323.

Rialbowol K, Fink S, Gilman M, Walsh D, Gooman R, and Feramisco JR (1998) The catalytic subunit of cAMP-dependet protein kinase induces expression of genes containing cAMP-responsive enhancer elements. *Nature* 336: 83–86.

Stacey DW (1981) Microinjection of mRNA and other macromolecules into living cells. *Methods Enzymol.* 79: 76–88.

The Use of Plasmid Microinjection to Study Specific Cell Cycle Phase Transitions

E.S. Knudsen

Contents

1   Introduction

1.1 Cell cycle summary

Progression through the eukaryotic cell cycle is a highly regulated process in which cellular proliferation is dependent on the integration of multiple signals (reviewed in, Hamel and Hanley-Hyde, 1997; Reed, 1997; Del Sal et al., 1996; Sherr, 1996; Palmero and Peters, 1996). Diverse extracellular signals such as growth factors, adhesion, and nutrient availability, as well as intracellular signals such as cell size and genomic integrity, must be monitored for proper proliferation. Mis-regulation of cellular proliferation, as occurs in cancer cells, can often be traced to mutations which influence cell cycle transitions, particularly those involved in the decision to engage in genome replication (Palmero and Peters, 1996; Hamel and Hanley-Hyde, 1997; Sherr, 1996; Hunter and Pines, 1995).

Methods and Tools in Biosciences and Medicine
Microinjection, ed. by J. C. Lacal et al.
© 1999 Birkhäuser Verlag Basel/Switzerland

Like all cell cycle transitions, the transition from G1 to S-phase is a strictly controlled process (Reed, 1997; Hamel and Hanley-Hyde; Del Sal et al., 1996). The activities of three sets of cyclin dependent kinase (cdk)/ cyclin complexes are ultimately responsive to environmental cues, which determine if the cellular environment favors DNA-replication. The G1 cdk/cyclin complexes consist of cyclin D, cyclin E, or cyclin A and their associated cdk catalytic partners, cdk4 and cdk2. All five proteins are required for progression from G1 to S-phase, as reagents which inhibit their activity arrest cells in G1 (Lukas et al., 1994; Ohtsubo et al., 1995; Pagano et al., 1992; Tsai et al., 1993); conversely, ectopic expression of these proteins can promote entry into S-phase (Resnitzky et al., 1994). The activity of cdk/cyclin complexes is influenced by regulated expression, protein degradation, phosphorylation, and complex formation with cdk-inhibitors (Reed, 1997; Del Sal et al., 1996; King et al., 1996; Sherr, 1996). Together, the interplay of these regulatory mechanisms determines the status of cdk/cyclin activity within the cell. For example, growth factors promote progression through G1 to S-phase by stimulating cyclin D expression (Sherr, 1994). Conversely, anti-mitogenic factors such as DNA-damage or TGF-β can act by inducing inhibitors of cdk/cyclin activity, such as p21cip1 and p15ink4b, respectively (Elledge, 1996; Hannon and Beach, 1994).

A critical substrate of G1 cdk/cyclin activity is the retinoblastoma tumor suppressor protein, RB (Sidle et al., 1996; Bartek et al., 1996; Palmero and Peters, 1996; Hamel and Hanley-Hyde, 1996; Wang et al., 1994). Loss of RB leads to the development of specific human tumor types, and an overall deregulation of the G1/S transition in cultured cells (Kaelin et al., 1997; Herrera et al., 1996; Wang et al., 1994; Hamel et al., 1993). Consistent with a role in tumor suppression, ectopic expression of RB in specific subsets of tumor cells inhibits cell growth or leads to reduced tumorigenicity (Wang et al., 1994; Hamel et al., 1993). RB functions in the cell as a "protein-binding-protein", binding to and regulating the activity of various cellular proteins (Sidle et al., 1996; Wang et al., 1994). This protein-binding activity is critical for RB biological function, as mutations which compromise RB protein-binding activity are associated with a lack of growth inhibitory activity and tumorigenesis (Otterson et al., 1997; Bremner et al., 1997).

Phosphorylation of RB is initiated in early to mid-G1 by cdk4/cyclin D complexes, and is maintained throughout the cell cycle through the action of other cdk/cyclin complexes until RB is rapidly dephosphorylated as cells reenter G1 from mitosis (Sherr, 1996; Sidle et al., 1996; Beijersbergen et al., 1996; Wang et al., 1994). Cdk/cyclin-mediated phosphorylation disrupts RB protein-binding activity and as such disrupts the ability of RB to inhibit entry into S-phase (Knudsen and Wang, 1997; Connell-Crowley et al., 1997; Knudsen and Wang, 1996; Hinds et al., 1992). Together, the cdk/cyclins and RB protein form a pathway which is responsive to the cellular environment to allow for proper DNA-replication, with cdk/cyclin activity impinging upon RB to allow for entry into S-phase (Sherr, 1996; Bartek et al., 1996; Hamel and Hanley-Hyde, 1996). The importance of RB as a cdk/cyclin substrate is underscored by the finding that cdk4 activity is no longer required for cell cycle progression in cells which are RB-deficient (Lukas et al., 1994; Lukas et al., 1995; Koh et al., 1995). Furthermore, the cdk/RB pathway is mis-regulated in numerous tumor types, indicating its importance in the regulated growth of cells (Hamel and Hanley-Hyde, 1996; Del Sal et al., 1996; Sherr et al., 1996; Bartek et al., 1996; Palmero and Peters, 1996).

Recently, several groups generated RB phosphorylation site mutant proteins which are refractory to cdk/cyclin-mediated phosphorylation, and demonstrate enhanced growth inhibitory activity (Knudsen and Wang, 1997; Lukas et al., 1997; Leng et al., 1997). The result of this growth inhibitory activity is similar to that induced by the cdk-inhibitors p16Ink4a, p21Cip1, and p27Kip1, as individual ectopic expression of any of these proteins has been shown to cause cells to accumulate in G1.

A key point in progression through the G1/S cell cycle transition is the "restriction point". This is defined as the point in G1 when cells no longer require extracellular stimuli, such as growth factor, for progression into and through S-phase (Del Sal et al., 1996; Sherr, 1996). The growth inhibitory action of RB and certain certain cdk-inhibitors has been hypothesized to function only before the restriction point (Del Sal et al., 1996; Sherr, 1996). To test this hypothesis, we investigated the role of cdk-inhibitors and constitutively active RB proteins in two specific portions of the cell cycle: i) the pre-restriction point early G1 to S-phase transition and ii) progression through S-phase which is post-restriction point.

1.2   Microinjection in the study of mammalian cell cycle regulation

As briefly discussed above, the regulation of cell cycle progression has been studied in great detail. Cultured mammalian cells provide a good system for studying the intricacies of mammalian cell cycle control. They are amenable to classical biochemical studies of cell cycle regulatory proteins, such as cdks and cyclins. Furthermore, plasmids can be readily introduced into many cultured cell types through transfection or retroviral transduction, allowing one to investigate the role proteins play in controlling cellular processes.

Additionally, microinjection has also been used extensively in the study of cell cycle regulation in mammalian cells. Microinjection of antibodies directed at cyclins have helped define their pivotal role in cellular proliferation. For example, microinjection of cyclin A antibodies into early G1-phase cells blocks S-phase entry, indicating that cyclin A is required for entry into S-phase (Pagano et al., 1992). Furthermore, microinjection of cyclin D1 antibodies into both RB-positive and RB-negative cell types showed that functional cyclin D1 is only required in cells containing functional RB (Lukas et al., 1994). Similarly, microinjection of both expression and reporter plasmids has been utilized to probe the action of specific proteins in cell cycle control and transcriptional events relating to cell cycle progression (Lukas et al., 1997). One of the benefits of using plasmids as opposed to purified proteins is their relative ease of preparation. Furthermore, with an ever increasing number of dominant negative proteins available, it is possible to carry out inactivating experiments without relying on antibodies. However, one of the disadvantages of using plasmid injection is the lag time between injection and the accumulation of the plasmid-encoded proteins. This is particularly problematic when studying transient events such as passage through cell cycle transitions.

An immediate benefit of microinjection in cultured mammalian cells is its use in cells which are difficult to transfect or infect, such as post-proliferative cells e.g. senescent or differentiated cells (Puri et al., 1996; Rose et al., 1992; Thorburn et al., 19993; Wadhwa et al., 1993). Furthermore, reagents can be delivered by microinjection in any phase of the cell cycle, which can be readily controlled through differential culturing of the cells. As such, microinjection has been particularly useful for determining the specific point(s) at which proteins act in cell cycle progression (Pagano et al., 1992; Goodrich et al., 1993). Here, we describe the use of two different techniques to investigate the activity of proteins on discrete cell cycle transitions using plasmid microinjection.

2   Materials

Chemicals

The plasmid encoding p16ink4a has been previously described (Knudsen and Wang, 1997; Knudsen et al., 1997).

Anti-Bromo-deoxy-Uridine (BrdU) antibodies	Accurate Scientific
Aphidicholin	Calbiochem
GFP expression plasmid	Gibco BRL
Green Lantern	Gibco BRL
Rhodamine labeled anti-Rat antibody	Jackson Immuno-Research Laboratories

3   Methods

3.1 Plasmid preparation

All plasmids utilized for micro-injection were doubly purified to remove any impurities. Plasmids were originally purified either by cesium chloride banding or using Qiagen Maxi-Prep kits. Fifty µg of the resulting plasmid DNA was then re-purified over a Qiagen Midi-column. Plasmid concentration was determined and integrity of all plasmids was determined by visual analysis following agarose gel electrophoresis. Plasmids were dissolved in micro-injection buffer as described for the given experiment.

3.2   Cell culture

Rat-1 immortalized rodent fibroblasts were obtained from Dr. Doug Green (La Jolla Institute for Allergy and Immunology). Cells were grown in DMEM supplemented with 10% fetal bovine serum (FBS) and antibiotics. For synchronization in quiescence, cells were cultured for 72 h in defined minimal media (DMM). For synchronization in early S-phase, cells were cultured in the presence of 2 µg/ml aphidicholin for 24 h. For microinjection, cells were plated on acid washed glass cover slips at least 48 h prior to microinjection.

3.3   Microinjection and immunofluorescence

Cells were microinjected with glass capillary needles, made using a Kopf vertical pipet puller. The plasmids were co-injected typically at the concentration of 100 ng/µl of GFP and 50 ng/µl of the effector plasmid. Injections were carried out using an Eppendorf microinjector. We co-injected the GFP-expression plasmid as the marker for injected cells rather than IgG since it is a better prognosticator of a productive injection giving rise to the production of plasmid encoded protein.

For the injection of S-phase cells, Rat-1 cells which had been cultured for 24 h in aphidicholin containing media were used (Fig. 1A). The injected cells were cultured for an additional 16 h in aphidicholin to allow the injected plasmids to express protein. Cells were then washed extensively to remove the aphidicholin, and BrdU was added to the media. Cells were then fixed following a 6h labeling period.

For the injection of early G1 cells, Rat-1 cells which had been rendered quiescent by culture in DMM for 72 h were released into media containing 10% FBS for 2 h and then injected (Fig. 1B). BrdU was added immediately following injection. Since the start of S-phase is approximately 15 h after release from quiescence, this gave approximately 13 h to establish expression of the injected plasmids prior to S-phase. Cells were then fixed after 18 h in the presence of BrdU.

Following microinjection and labeling with BrdU to measure progression through S-phase, cells were fixed in 3.7% formaldehyde in PBS for 15 min and then permeabilized in 0.3% Triton X-100 in PBS for 10 min. Primary antibody staining (1:500 dilution) was carried out in PBS supplemented with 5mg/ml BSA and 0.5%NP-40 for 1 h at 37°C. in a humidified chamber. The cells were then washed in PBS. Secondary antibody (1:100 dilution) was diluted and incu-

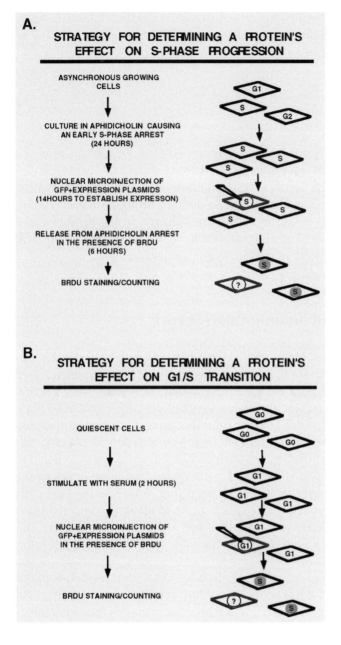

A.

STRATEGY FOR DETERMINING A PROTEIN'S
EFFECT ON S-PHASE PROGRESSION

ASYNCHRONOUS GROWING
CELLS

CULTURE IN APHIDICHOLIN CAUSING
AN EARLY S-PHASE ARREST
(24 HOURS)

NUCLEAR MICROINJECTION OF
GFP+EXPRESSION PLASMIDS
(14HOURS TO ESTABLISH EXPRESSON)

RELEASE FROM APHIDICHOLIN ARREST
IN THE PRESENCE OF BRDU
(6 HOURS)

BRDU STAINING/COUNTING

Figure 1A Microinjection strategy for S-phase progression.

Asynchronously growing Rat-1 cells were arrested in S-phase using the DNA-synthesis inhibitor aphidicholin. Arrested cells were microinjected and retained in the presence of aphidicholin for an additional 16 h to allow for the expression of the plasmid encoded gene products. Aphidicholin was removed by washing the cells and BrdU was added in fresh media for a 6h labeling period. Progression through S-phase was then monitored by BrdU-incorporation.

B.

STRATEGY FOR DETERMINING A PROTEIN'S
EFFECT ON G1/S TRANSITION

QUIESCENT CELLS

STIMULATE WITH SERUM (2 HOURS)

NUCLEAR MICROINJECTION OF
GFP+EXPRESSION PLASMIDS
IN THE PRESENCE OF BRDU

BRDU STAINING/COUNTING

Figure 1B Microinjection strategy for G1/S progression.

Rat-1 cells were rendered quiescent *via* culture in serum-free media for 72 h. These cells were then stimulated with media containing 10% fetal bovine serum for 2 h and then microinjected in early G1 at which time BrdU was added. This lag allowed the the cells approximately 13 h to establish protein expression prior to entry into S-phase. Progression through S-phase was monitored by BrdU-incorporation after 18 h of BrdU labeling.

bated as described for the primary antibody. Cells were washed again, mounted on glass coverslips and visualized by fluorescence microscopy.

Fluorescent microscopy was performed with a Zeiss Axiophot Microscope using either a 40× or 63× lens. Photographs of stained cells were recorded using a Hamamatsu CCD camera. Digital images were printed using a Mitsubishi color sublimation printer.

3.4 Controls

Several important controls were carried out prior to any of the experiments in-volving microinjection. First, we determined that the cells could be reversibly arrested in specific phases of the cell cycle, using FACS analysis and BrdU in-corporation of the uninjected Rat-1 cells. Specifically, we showed that the quiescent Rat-1 cells could be induced to synchronously enter S-phase within 15 h of the addition of growth media. We also showed that aphidicholin ar-rested the Rat-1 cells in early S-phase, without significant cell death. Release of these cells by washing the plates extensively allowed immediate and syn-chronous progression through S-phase into G2.

It has been previously shown that microinjection into early G1 cells sup-ports the expression of plasmid-encoded genes. However, it was unclear whether plasmid-encoded proteins would actually be expressed in the aphidi-cholin arrested cells. To test this, aphidicholin arrested cells were injected with a GFP-expression plasmid. The expression of GFP from the plasmid was as-sessed by immunofluorescence microscopy on the live cells over a period of 8 h. By 5 h, GFP-positive cells were readily visible, indicating that aphidicholin would not prevent the expression of plasmid borne genes.

A final control was carried out to demonstrate that simple microinjection of plasmid DNA into these cells did not cause a non-specific arrest of the cells in either G1 or S-phase. To test this, parental vector purified under identical con-dition as the test plasmids was assayed for its influence on cell cycle progres-sion. In our system, injection of parental plasmid DNA was never shown to in-fluence cell cycle progression.

4   Procedures and Applications

We were interested in studying the activity of cdk-inhibitors and a constitu-tively active RB protein at two specific phases of the cell cycle which could each be monitored by BrdU incorporation. Specifically, we set out to analyze the ef-fect of each protein on the G1 to S-phase transition (Fig. 1B), and progression through S-phase (Fig. 1A).

4.1 Cdk-inhibitors and constitutively active RB proteins inhibit S-phase entry from early G1

Rat-1 cells were initially synchronized in quiescence by serum deprivation. Ad-dition of 10% fetal bovine serum allows these cells to synchronously proceed into S-phase approximately 15 h after growth factor addition. Two h after ser-um addition, cells were injected with either vector or various effector expres-

A.

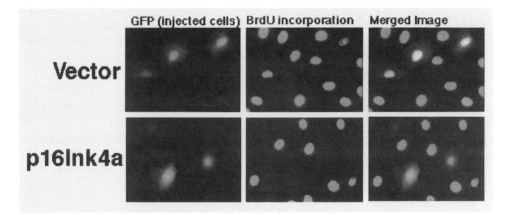

Figure 2A   Early G1 cells are serum dependent for entry into S-phase.

Rat-1 cells were rendered quiescent by culture in serum-free media for 72 h. Fresh media containing 10% fetal bovine serum was added to the quiescent cells for 2 h. Cells were then washed extensively with PBS and placed in media containing either 0% or 10% fetal bovine serum. BrdU was added immediately to the cells which were fixed after 18 h of labeling. BrdU-incorporation was measured by immunofluorescence staining with anti-BrdU antibody. Data shown is from two independent experiments with at least 300 cells counted per experiment.

Figure 2B   p16ink4a inhibits the G1/S transition.

Rat-1 cells were made quiescent by culture in serum-free media for 72 h. These cells were then stimulated with media containing 10% fetal bovine serum, and after 2 h subjected to microinjection with plasmids encoding GFP, and either vector control or p16ink4a. BrdU was added to the cells immediately following injection, and cells were fixed and processed for BrdU-incorporation after 18 h of labeling. Cells were stained for BrdU incorporation, and representative photographs taken at 63× magnification are shown.

sion plasmids, such as p16Ink4a. Two h following release from quiescence, the cells are observed to *not* be past the serum restriction point. This was evident in that if these cells were washed and placed in serum free media, there was little BrdU incorporation as compared with cells cultured in media containing 10% fetal serum (Fig. 2A).

Since cellular damage can non-specifically cause an arrest of these cells, it was prudent to verify that the mere injection of plasmid DNA did not cause a cell cycle arrest. Microinjection of vector and GFP had no effect on cell cycle progression, and greater than 80% of the injected cells (GFP-positive) registered BrdU-positive (Fig. 2B). These results indicate that under these experimental conditions, microinjection did not influence cell cycle progression into S-phase.

Injections of cdk-inhibitors into cells synchronized in early G1 were then monitored using the same assay conditions. We found that p16Ink4a inhibited progression from early G1 to S-phase (Fig. 2B). This is consistent with a number of published reports, wherein ectopic expression of p16Ink4a specifically arrests cells in G1 (Lukas et al., 1997). We also found that other cdk-inhibitors (p27Kip1 and p21Cip1) and constitutively active RB (PSM.7-LP) expression plasmids could also inhibit entry into S-phase. These results confirm that a specific arrest in G1 is induced by all of these proteins to prevent entry into S-phase, and indicate that these proteins can indeed arrest entry into S-phase, when introduced prior to the restriction point. For a full description of our results see Knudsen et al., 1998.

4.2   Method for the study of S-phase progression

The growth inhibitory action of specific cdk-inhibitors and RB are believed to be manifested only in the G1 phase of the cell cycle, prior to the restriction point. To determine whether any of these proteins could influence cell cycle progression after restriction point, we assessed the effect of expressing these proteins in aphidicholin arrested cells by microinjection. Aphidicholin arrests cells in early S-phase by inhibiting the activity of DNA-polymerase. This point is after the restriction point, as cells arrested in aphidicholin are able to progress through S-phase in the absence of serum (Fig. 3A). Aphidicholin blocked cells were found to be capable of expressing the plasmid-encoded proteins, as indicated by the accumulation of GFP in the cells following injection (Fig. 3B). We also found that microinjection of the GFP expression plasmid had no influence on BrdU incorporation following release from the aphidicholin block (Fig. 3B), indicating that the injection of plasmid DNA does not non-specifically inhibit DNA-synthesis. Using this approach we found that PSM-RB inhibits progression through S-phase (Knudsen et al., 1998).

5   Remarks and Conclusions

Microinjection has been effectively utilized in the study of cell cycle regulation. Here we focused on studying specific phase transitions. The passage from G1 to S-phase has been extensively studied. Predominantly, these types of studies have been carried out through the transfection of expression plasmids which

A.

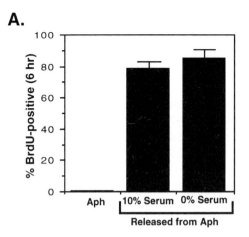

Figure 3A Aphidicholin blocked cells are reversibly arrested and serum independent for S-phase progression.

Rat-1 cells arrested in S-phase for 24 h with the DNA-polymerase inhibitor aphidicholin were washed extensively with serum-free media. BrdU was then added to the cells which were cultured in the presence or absence of fetal bovine serum. Cells were fixed and stained for BrdU incorporation following a 6h labeling period. The percentage of cells with positive BrdU staining is shown. Data is from two independent experiments with at least 300 cells counted per experiment.

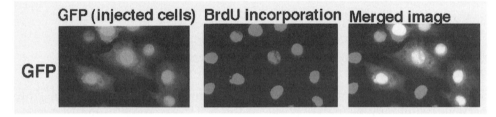

Figure 3B   Aphidicholin blocked cells express plasmid encoded proteins.

Rat-1 cells were arrested in S-phase following 24 h of culture in the presence of aphidicholin. These cells were microinjected with a GFP expression plasmid (100ng/μl). Cells were cultured for an additional 16 h in the presence of aphidicholin to allow for the accumulation of plasmid encoded proteins. These cells were then washed extensively, and cultured in fresh media containing BrdU for 6 h. The cells were then fixed and stained for BrdU-incorporation. Representative photographs were taken at 63× magnification.

ultimately lead to an increase in the G1 phase of the cell cycle. Using the micro-injection approaches discussed, we showed that cdk-inhibitors and constitutively active RB act specifically in G1 to inhibit S-phase entry of the injected cells. This result is consistent with similar experiments carried out by others (Lukas et al., 1997), and has been published (Knudsen et al., 1998). Overall, these results show that these proteins are capable of preventing progression into S-phase through specific actions carried out during the G1 phase of the cell cycle.

Since the regulation of S-phase progression remains largely unstudied, we established a method to specifically assay for the role of individual proteins in S-phase progression. This assay relies on arresting cells in S-phase using the DNA polymerase inhibitor aphidicholin. A stable arrest of the cells is important since it enables accumulation of the plamid-encoded protein. Aphidicholin also allowed for injection into cells which were stably yet reversibly arrested, allowing accumulation of the plasmid-encoded proteins prior to release. As demonstrated in this report, aphidicholin treated cells are arrested post-restriction point, enabling us to assess the role of individual proteins on S-phase progression independent of effects on G1-traversal. This work and additional results have been published elsewhere (Knudsen et al., 1998).

The use of techniques, such as those outlined here, should prove useful in probing the role of proteins in specific cell cycle transitions. Like a number of physiological processes, the study of S-phase progression is hindered by its transient nature. However, by using drugs (aphidicholin) which reversibly block S-phase progression, we are able to study the action of plasmid expressed proteins on S-phase. Similar approaches using microinjection in conjunction with agents which block other cell cycle phases, such as the completion of mitosis, should prove useful in future studies.

Acknowledgements

Erik S. Knudsen worked in the laboratory of James R. Feramisco, where he was supported by grants to JRF from California Tobacco Related Diseases Research Program and a training grant from the NIH to the UCSD Cancer Center (T32CA09290). The author is thankful to James R. Feramisco and members of the laboratory for thought provoking discussion, and technical assistance, particularly Carolan Buckmaster for training in microinjection. Furthermore, Jean Y.J. Wang's advice was instrumental in the implementation of the techniques investigating S-phase progression. The author is also indebted to Karen E. Knudsen for critical discussion.

References

Bartek J, Bartkova J, Lukas J (1996) The retinoblastoma protein pathway and the restriction point. Current Op Cell Biol 8(6): 805–814.

Beijersbergen RL, Bernards R (1996) Cell cycle regulation by the retinoblastoma family of growth inhibitory proteins. Biochim Biophys Acta 1287(2–3): 103–120.

Bremner R, Du DC, Connolly-Wilson MJ, Bridge P, Ahmad KF, Mostachfi H, Rushlow D, Dunn JM, Gallie BL (1997) Deletion of RB exons 24 and 25 causes low-penetrance retinoblastoma. Am J Human Genetics 61: 556–570.

Connell-Crowley L, Harper JW, Goodrich DW (1997) Cyclin D1/Cdk4 regulates retinoblastoma protein-mediated cell cycle arrest by site-specific phosphorylation. Molecular Biol Cell 8: 287–301.

DelSal, G, Loda, M, Pagano, M (1996) Cell cycle and cancer: Critical events at the G1 restriction point. Crit. Rev. Onc. V7 N1–2: 127–142.

Elledge SJ (1996) Cell cycle checkpoints: preventing an identity crisis. Science 274: 1664–1672.

Goodrich DW, Wang NP, Qian YW, Lee EY, Lee WH (1991) The retinoblastoma gene product regulates progression through the G1 phase of the cell cycle. Cell 67: 293–302.

Hannon GJ, Beach D (1994) p15INK4B is a potential effector of TGF-beta-induced cell cycle arrest. Nature 371: 257–261.

Hamel PA, Hanley-Hyde J (1997) G1 cyclins and control of the cell division cycle in normal and transformed cells. Cancer Investigation 2: 143–152.

Hamel PA, Phillips RA, Muncaster M, Gallie BL (1993) Speculations on the roles of RB1 in tissue-specific differentiation, tumor initiation, and tumor progression. Faseb Journal 10: 846–854.

Herrera RE, Sah VP, Williams BO, Makela TP, Weinberg RA, Jacks T (1996) Altered cell cycle kinetics, gene expression, and G1 restriction point regulation in Rb-deficient fibroblasts. Molecular Cell Biol 5: 2402–2407.

Hinds PW, Mittnacht S, Dulic V, Arnold A, Reed SI, Weinberg RA (1992) Regulation of retinoblastoma protein functions by ectopic expression of human cyclins. Cell 70: 993–1006.

Hunter T, Pines J (1994) Cyclins and cancer. II: Cyclin D and CDK inhibitors come of age. Cell 79: 573–582.

Kaelin WG Jr (1997) Recent insights into the functions of the retinoblastoma susceptibility gene product. Cancer Investigation 15: 243–254.

Kaul SC, Mitsui Y, Komatsu Y, Reddel RR, Wadhwa R (1996) A highly expressed 81 kDa protein in immortalized mouse fibroblast: its proliferative function and identity with ezrin. Oncogene 13: 1231–1237.

King RW, Deshaies RJ, Peters JM, Kirschner MW (1996) How proteolysis drives the cell cycle. Science 274: 1652–1659.

Knudsen ES, Wang JY (1997) Dual mechanisms for the inhibition of E2F binding to RB by cyclin-dependent kinase-mediated RB phosphorylation. Molecular Cell Biol 17: 5771–5783.

Knudsen ES, Chen TT, Buckmaster C, Feramisco JR, Wang JY (1998) Inhibition of DNA synthesis by RB: distinct effect on the G1/S transition and S-phase progression. Genes and Development 12: 2278–2292.

Koh J, Enders GH, Dynlacht BD, Harlow E (1995) Tumour-derived p16 alleles encoding proteins defective in cell-cycle inhibition. Nature 375: 506–510.

Leng XH, Connell Crowley L, Goodrich D, Harper JW (1997) S-phase entry upon ectopic expression of G1 cyclin-dependent kinases in the absence of retinoblastoma protein phosphorylation. Current Biology, V7 N9: 709–712.

Lukas J, Parry D, Aagaard L, Mann DJ, Bartkova J, Strauss M, Peters G, Bartek J (1995) Retinoblastoma-protein-dependent cell-cycle inhibition by the tumour suppressor p16. Nature 375: 503–506.

Lukas J, Muller H, Bartkova J, Spitkovsky D, Kjerulff AA, Jansen-Durr P, Strauss M, Bartek J (1994) DNA tumor virus oncoproteins and retinoblastoma gene mutations share the ability to relieve the cell's requirement for cyclin D1 function in G1. J Cell Biology 125: 625–638.

Lukas J, Herzinger T, Hansen K, Moroni MC, Resnitzky D, Helin K, Reed SI, Bartek J (1997) Cyclin E-induced S phase without activation of the pRb/E2F pathway. Genes Development 11: 1479–1492.

Ohtsubo M, Theodoras AM, Schumacher J, Roberts JM, Pagano M (1995) Human cyclin E, a nuclear protein essential for the G1-to-S phase transition. Molecular Cell Biology 15: 2612–2624.

Otterson, GA, Chen, WD, Coxon, AB, Khleif, SN, Kaye FJ (1997) Incomplete penetrance of familial retinoblastoma linked to germ-line mutations that result in partial loss of RB function. Proc Natl Acad Sci USA V94 N22: 12036–12040.

Pagano M, Pepperkok R, Verde F, Ansorge W, Draetta G (1992) Cyclin A is required at two points in the human cell cycle. Embo J 11: 961–971.

Palmero I, Peters G (1996) Perturbation of cell cycle regulators in human cancer. Cancer Surveys 27: 351–367.

Puri PL, Avantaggiati ML, Balsano C, Sang N, Graessmann A, Giordano A, Levrero M (1997) p300 is required for MyoD-dependent cell cycle arrest and muscle-specific gene transcription. Embo J 16: 369–383.

Reed SI (1997) Control of the G(1)/S transition. Cancer Surveys V29: 7–23.

Resnitzky D, Gossen M, Bujard H, Reed SI (1994) Acceleration of the G1/S phase transition by expression of cyclins D1 and E with an inducible system. Molecular and Cellular Biology 14: 1669–1679.

Rose DW, McCabe G, Feramisco JR, Adler M (1992) Expression of c-fos and AP-1 activity in senescent human fibroblasts is not sufficient for DNA synthesis. J. Cell Biology 268: 1405–1411.

Sherr CJ (1994) G1 phase progression: cycling on cue. Cell 79: 551–555.

Sherr CJ (1996) Cancer cell cycles. Science 274: 1672–1677.

Sidle A, Palaty C, Dirks P, Wiggan O, Kiess M, Gill RM, Wong AK, Hamel PA (1996) Activity of the retinoblastoma family proteins, pRB, p107, and p130, during cellular proliferation and differentiation. Crit Rev Biochem Molecular Biol 31: 237–271.

Thorburn A, Thorburn J, Chen SY, Powers S, Shubeita HE, Feramisco JR, Chien KR (1993) Hras dependent pathway for cardiac muscle hypertrophy. J Biol Chem 268: 2244–2249.

Tsai LH, Lees E, Faha B, Harlow E, Riabowol K (1993) The cdk2 kinase is required for the G1-to-S transition in mammalian cells. Oncogene 8: 1593–1602.

Wang JY, Knudsen ES, Welch PJ (1994) The retinoblastoma tumor suppressor protein. Advances Cancer Res 64: 25–85.

Combining Microinjection with Microanalytical Methods to Investigate Regulation of Metabolic Pathways

J. Manchester and J.C. Lawrence, Jr.

Contents

1   Introduction

Microinjection is a powerful tool for studying the regulation of cellular function. It offers a distinct advantage over other techniques for investigating signal transduction pathways involved in acute hormonal responses, such as those contributing to the effect of insulin on lowering of blood glucose. This important response to insulin involves the stimulation of glucose transport into muscle and fat cells, and the storage of the glucose as glycogen. The stimulatory effect on glycogen synthesis reaches a maximum within minutes of exposing a sensitive cell to the insulin. Microinjection provides a means for changing the intracellular concentrations of putative signaling intermediates within this time frame. By comparison, changes effected by gene transfections typically require hours to days, and can not be assumed to be representative of the acute metabolic response. The potential of microinjection in the investigation of metabolic responses in cells has not been realized, in part because of the perception that methods having the necessary sensitivity are not available. Actually, the capability has existed for many years (Passonneau and Lowry, 1992). The microanalytical methodology developed by the late Oliver Lowry provides a basis for measuring directly, and with high precision, enzyme activities and metabolites from single cells. In this chapter we describe protocols for

Methods and Tools in Biosciences and Medicine
Microinjection, ed. by J. C. Lacal et al.
© 1999 Birkhäuser Verlag Basel/Switzerland

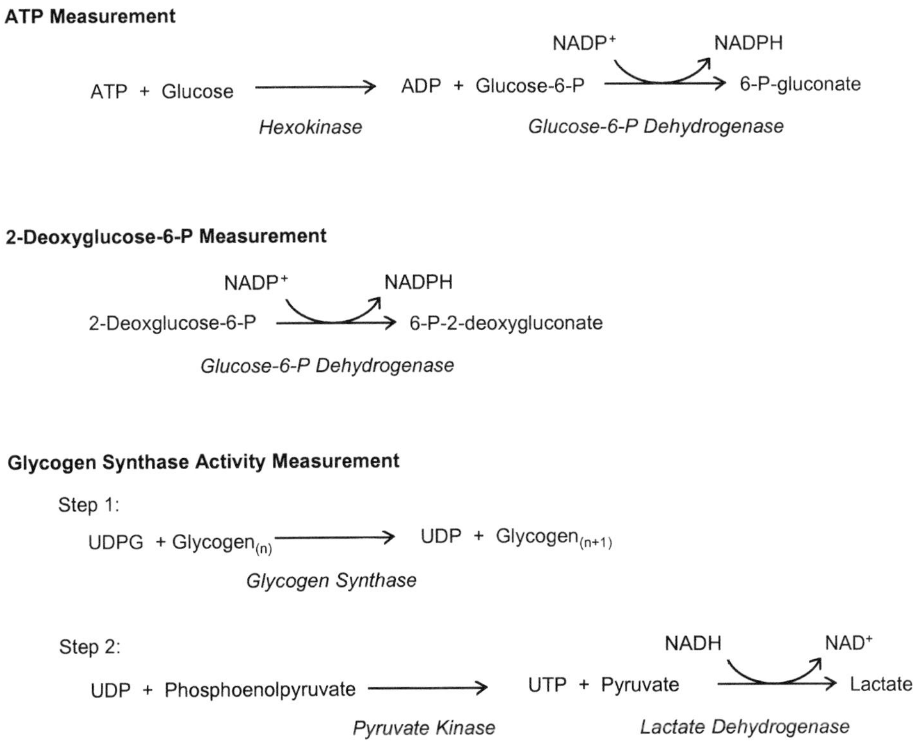

ATP Measurement

$$\text{ATP + Glucose} \xrightarrow{\text{Hexokinase}} \text{ADP + Glucose-6-P} \xrightarrow[\text{Glucose-6-P Dehydrogenase}]{\text{NADP}^+ \searrow \nearrow \text{NADPH}} \text{6-P-gluconate}$$

2-Deoxyglucose-6-P Measurement

$$\text{2-Deoxglucose-6-P} \xrightarrow[\text{Glucose-6-P Dehydrogenase}]{\text{NADP}^+ \searrow \nearrow \text{NADPH}} \text{6-P-2-deoxygluconate}$$

Glycogen Synthase Activity Measurement

Step 1:

$$\text{UDPG + Glycogen}_{(n)} \xrightarrow{\text{Glycogen Synthase}} \text{UDP + Glycogen}_{(n+1)}$$

Step 2:

$$\text{UDP + Phosphoenolpyruvate} \xrightarrow{\text{Pyruvate Kinase}} \text{UTP + Pyruvate} \xrightarrow[\text{Lactate Dehydrogenase}]{\text{NADH} \searrow \nearrow \text{NAD}^+} \text{Lactate}$$

Figure 1 The specific reactions involved in measuring levels of ATP and 2-deoxyglucose-6-P, and the activity of glycogen synthase.

Note that the reactions are coupled to pyridine nucleotide oxidation or reduction.

using microinjection to investigate processes involved in the regulation of glycogen synthesis in cardiac myocytes and 3T3-L1 adipocytes, two insulin-responsive cell types.

Levels of enzymes and metabolites from single cells are measured in a series of reactions coupled to pyridine nucleotide oxidation or reduction (Fig. 1). There is nothing unusual about the concentrations of auxiliary enzymes, cofactors, or substrates used in these reactions; however, it is essential that the reactions be conducted in small volumes to minimize the reagent blank caused by contaminants that are inevitably present in the enzymes and cofactors used. For this reason, reactions are conducted under oil to prevent evaporation and dispersion of the droplet. The incubations are terminated by adding either acid or base (see below), and heating the samples. Amplification of the signal is accomplished by enzymatically cycling the pyridine nucleotide product formed (Fig. 2).

The concept of enzymatic cycling to amplify signals is now widely appreciated as a result of the use of the polymerase chain reaction to amplify DNA. Amplification by pyridine nucleotide cycling is equally elegant in its simplicity.

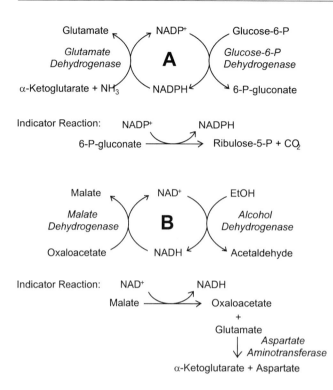

Figure 2 Pyridine nucleotide cycles.

In the NADP (A) and NAD (B) cycles the pyridine nucleotide products formed in the specific reactions are cycled between oxidized and reduced forms by using opposing dehydrogenases and the appropriate substrates. Note that the cycles do not discriminate between oxidized or reduced nucleotides. Therefore, it is necessary to destroy the NADP and the NADH used as substrates in the specific reactions before performing the cycling step. This is possible because the oxidized and reduced forms of the nucleotides are differentially sensitive to acid and base. Indicator reactions are used to measure the products of the cycling reactions.

Depending on the assay, either the NAD or the NADP amplification cycle is used (Passonneau and Lowry, 1992). The same principles are involved in both. For the NAD cycle shown in Fig. 2, a reaction mixture is prepared containing the opposing dehydrogenases, malate dehydrogenase and alcohol dehydrogenase, and saturating concentrations of the respective substrates, oxaloacetate and ethanol. Until pyridine nucleotide is added, nothing happens. However, with the addition of either NADH or NAD, the nucleotide is cycled between the oxidized and reduced form. Each turn of the cycle generates malate and represents a fold amplification. Because adding either NAD or NADH will result in cycling, it is necessary to destroy the substrate before proceeding to the cycling reaction. Fortunately, this is easily accomplished as the oxidized and reduced forms of the pyridine nucleotide are differentially sensitive to acid and base (Passonneau and Lowry, 1992). For example, NADH is destroyed by HCl at concentrations that have little effect on NAD. After sufficient amplification, the cycling reactions are terminated and the amounts of malate are measured in a final indicator reaction (Fig. 2). The cycling rate can be predicted from the kinetic constants of the two dehydrogenases, but it is measured directly in every experiment by including the appropriate pyridine nucleotide standards. In the assays of glycogen synthase activity and levels of ATP and 2-deoxyglucose-6-P, the NADPH or NADH formed in the final step can be measured using a benchtop fluorometer. The signal amplification necessary differs, depending on the assay.

2   Materials

All of the chemicals and auxiliary enzymes used in the assays described are commercially available. However, some of the equipment used in microanalysis is highly specialized and must be hand-made. A detailed description of the micropipets, fish pole balances, and other microanalytical tools is beyond the scope of this chapter. The reader is referred to the text of Passonneau and Lowry (1992) for more complete descriptions of these instruments and their construction.

Chemicals

Alcohol dehydrogenase (EC 1.1.1.1)	Sigma	A 3263
ADP	Sigma	A 6521
Ammonium acetate	Sigma	A 7262
AMP Buffer	Sigma	A 9199
Aspartate amino transferase (2.6.1.1)	Boehringer Mannheim	105 554
ATP	Sigma	A 5394
Bovine serum albumin	Sigma	A 6003
Catalase (EC 1.11.1.6)	Boehringer Mannheim	106 836
2-Deoxyglucose	Sigma	D 6134
2-Deoxyglucose-6-P	Sigma	D 8875
Dithiothreitol	Sigma	D 5545
EDTA	Sigma	EDS
Ethanol	Sigma	E 9878
Glucose	Sigma	G 7528
Glucose-6-P	Sigma	G 7250
Glucose-6-P dehydrogenase (EC 1.1.1.49) (from *Leuconostoc mesenteroides*)	Calbiochem	346774
Glutamate	Sigma	G 1626
Glutamate dehydrogenase (EC 1.4.1.3)	Boehringer Mannheim	127 701
Glycogen	Sigma	G 4011
Hexadecane	Sigma	H 0255
Hexokinase (EC 2.7.1.1)	Boehringer Mannheim	1 426 362
Hydrogen peroxide	Sigma	H 1009
Imidazole	Sigma	I 0250
Insulin	Eli Lily	
Potassium acetate	Sigma	P 1147
α-Ketoglutarate	Sigma	K 3752
KF	Sigma	P 2569
Lactate dehydrogenase (EC 1.1.1.27)	Boehringer Mannheim	106 992
Mineral oil, light	Sigma	M 3516
β-Mercaptoethanol	Sigma	M 6250
$MgCl_2$	Sigma	M 8266

Malate dehydrogenase (EC 1.1.1.37)	Sigma	M 7383
NAD	Sigma	N 6754
NADH	Sigma	N 6879
NADP	Sigma	N 5881
Oxaloacetate	Sigma	O 4126
Phosphoenol pyruvate	Sigma	P 7002
Phosphofructokinase (EC 2.7.1.11)	Sigma	F 6877
6-P-Gluconate	Sigma	P 6888
6-P-Gluconate dehydrogenase (EC 1.1.1.44)	Boehringer Mannheim	108 405
Phosphoglucose isomerase (EC 5.3.1.9)	Boehringer Mannheim	128 139
Pyruvate kinase (EC 2.7.1.40)	Boehringer Mannheim	128 155
Rhodamine dextran	Collaborative Research	
Tris	Sigma	T 6791
UDP	Sigma	U 4125
UDP-Glucose	Sigma	U 4625

Equipment

- Ace Glass vacuum tubes
- Aluminum tissue holders
- Borosilicate glass culture tubes
- Constriction pipets
- Dissecting room maintained at 20°C and 40–50% humidity
- Dissecting stages
- Dry temperature block
- Flaming/Brown micropipet puller
- Fishpole Balances
- Filter Flourometer (excitation 340 nm; emission 460 nm)
- Horizontal microscope with a micrometer ocular in one eyepiece
- Micro dissection tools and manipulators
- Oil Wells
- Wide-field microscope with light from below

Solutions

- **ATP Measurement Reagent (Solution 1)**
    100 mM Tris-acetate, pH 8.5
    2 mM $MgCl_2$
    0.04% bovine serum albumin
    0.2 mM glucose
    0.05 mM NADP
    4 µg/ml hexokinase
    1 µg/ml glucose-6-phosphate dehydrogenase

- **Glucose-6-P Removal Reagent (Solution 2)**
     100 mM Tris-acetate, pH 8.5
     2 mM $MgCl_2$
     0.04% bovine serum albumin
     100 mM potassium acetate
     2 mM phosphoenolpyruvate
     0.6 mM ATP
     20 µg/ml phosphoglucose isomerase
     32 µg/ml phosphofructokinase
     10 µg/ml pyruvate kinase
- **2-Deoxyglucose-6-P Measurement Reagent (Solution 3)**
     20 mM Tris-acetate, pH 8.1
     0.02% bovine serum albumin
     9 mM $MgCl_2$
     0.03 mM NADP
     150 µg/ml glucose-6-P dehydrogenase
- **Cycling Reagent (Solution 4)**
     100 mM imidazole-HCl, pH 7.0
     2 mM glucose-6-phosphate
     7.5 mM disodium α-ketoglutarate
     0.1 mM ADP
     25 mM ammonium acetate
     0.1% bovine serum albumin
     100 µg/ml glutamate dehydrogenase
     20 µg/ml glucose-6-phosphate dehydrogenase
- **Indicator Reagent (Solution 5)**
     100 mM Imidazole-HCl, pH 7.0
     30 mM ammonium acetate
     2 mM $MgCl_2$
     0.1 mM EDTA
     0.1 mM NADP
     2.5 µg/ml 6-P-gluconate dehydrogenase
- **Preincubation Reagent (Solution 6)**
     0.05% BSA
     5 mM EDTA
     2 mM glycogen
     0.5 mM dithiothreitol
     50 mM KF
     50 mM Imidazole-HCl, pH 7.4

- **Glycogen Synthase Reaction Mixture (Solution 7)**
    4 mM uridine 5'-diphosphoglucose
    0.05% bovine serum albumin
    5 mM EDTA
    2 mM glycogen
    0.5 mM dithiothreitol
    50 mM KF
    50 mM imidazole-HCl, pH 7.4
    (either minus glucose-6-P or plus 20mM glucose-6-P)
- **UDP Measurement Reagent (Solution 8)**
    200 mM imidazole-HCl, pH 6.7
    0.05% bovine serum albumin
    14 mM $MgCl_2$
    0.8 mM phosphoenolpyruvate
    100 mM potassium acetate
    40 µg/ml pyruvate kinase
    2 µg/ml lactate dehydrogenase (beef heart)
    4 µg/ml catalase
    40 µM NADH (for glycogen synthase activity minus glucose-6-P)
    240 µM NADH (for glycogen synthase activity plus glucose-6-P)
- **Cycling Reagent (Solution 9)**
    For glycogen synthase activity assayed in the absence of glucose-6-P):
    100 mM Tris-HCl, pH 8.1
    2 mM β-mercaptoethanol
    2 mM oxalacetate
    0.3 M ethanol
    0.02% bovine serum albumin
    100 µg/ml alcohol dehydrogenase
    10 µg/ml malate dehydrogenase
    For glycogen synthase activity assayed in the presence of glucose-6-P:
    Same as above except with 20 µg/ml alcohol dehydrogenase and
    2 µg/ml malate dehydrogenase.
- **Malate Indicator Reagent (Solution 10)**
    50 mM 2-amino-2 methylpropanol-HCl, pH 9.9
    5 µM glutamate, pH 9.9
    0.1 mM NAD
    5 µg/ml malate dehydrogenase
    2 µg/ml glutamate-oxaloacetate transaminase

3   Methods

To assess glucose transport, cells are incubated with 2-deoxyglucose, a glucose analog that is transported by the GLUT family of facilitative glucose transporters (Mueckler, 1994). 2-Deoxyglucose is phosphorylated by hexokinase, but further metabolism occurs to a very limited extent. Under appropriate conditions, essentially all of the intracellular 2-deoxyglucose is converted to 2-deoxyglucose-6-P, so that measuring the initial rate of accumulation of 2-deoxyglucose-6-P provides an index of glucose transport activity (Manchester et al., 1994a).

Glycogen synthase, the enzyme that catalyzes the incorporation of glucose from UDP-glucose into glycogen, is controlled by phosphorylation and by several allosteric effectors (Skurat and Roach, 1996). Phosphorylation decreases synthase activity if it is measured in the absence of glucose-6-P, but not if it is measured in the presence of high glucose-6-P, as even highly phosphorylated synthase is fully activated by glucose-6-P. Insulin promotes dephosphorylation of glycogen synthase producing an activation that persists in extracts, assuming care is taken to preserve the phosphorylation state of the enzyme. Thus, the hormonal effect may be monitored by the increase in the minus glucose-6-P/ plus glucose-6-P activity ratio.

The overall strategy is illustrated in Fig. 3. Cells are injected with the desired substances, then incubated with insulin or other agents, and with 2-deoxyglucose if glucose transport is being assessed. The incubations are terminated and the samples are freeze-dried at -35°C before myocytes are weighed (it is not practical to weigh adipocytes, discussed later). Single myocytes or clusters of adipocytes are transferred to oil wells where metabolite levels and enzyme activities are measured in reactions coupled to pyridine nucleotide oxidation or reduction.

3.1   Glucose transport after microinjection of cardiac myocytes

The procedures described below were developed for assessing glucose transport after injecting cardiac myocytes (Manchester et al., 1994b). Myocytes are isolated by collagenase digestion of adult rat ventricles and are seeded onto laminin-coated coverslip fragments (1–2 mm^2). All agents injected are dissolved in a buffer containing rhodamine-labeled dextran to allow identification of injected cells (Manchester et al., 1994b). Methods for microinjecting these cells are covered elsewhere in this text and will not be described in this chapter. After injection, cells are treated with insulin or other additions as desired. For assessing glucose transport activity after microinjection, the cells are incubated for 10 min in glucose-free medium supplemented with 9 mM lactate, 1 mM pyruvate, and 0.1 mM 2-deoxyglucose.

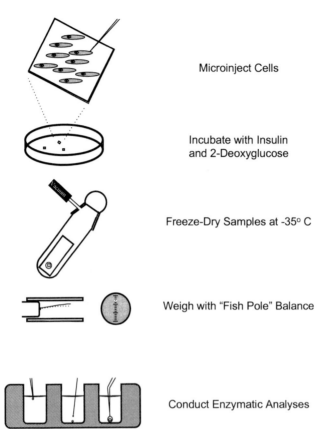

Figure 3   Steps in assessing metabolic reponses after microinjection.

Microinject Cells

Incubate with Insulin
and 2-Deoxyglucose

Freeze-Dry Samples at -35° C

Weigh with "Fish Pole" Balance

Conduct Enzymatic Analyses

Use of a low concentration of 2-deoxyglucose is important to ensure that the transport step is rate-limiting.

Cardiac myocytes are relatively large and are only loosely attached via the laminin matrix to the coverslips. These are desirable properties for the microanalytic work, as individual cells may be picked up and weighed with relative ease. A disadvantage is that adult myocytes are very susceptible to damage by microinjection. The number of viable cells can be increased by performing injections in the presence of butanedione monoxime (Manchester et al., 1994b). Nevertheless, not every cell survives, and a means for detecting and eliminating damaged cells is needed. ATP levels provide one criterion for assessing viability. Measuring ATP has proven very useful, as injected cells that appear normal by light microscopy may sometimes have only a small fraction of the normal ATP (Manchester et al., 1994b). Procedures for measuring both ATP and 2-deoxyglucose-6-P in single myocytes are described below.

Protocol 1   Measuring ATP and 2-deoxyglucose-6-P in single myocytes

1. Quick-freeze by placing the coverslip fragments with injected cells in aluminum holders prechilled on dry ice. Place the holders in pre-chilled vacuum
   tubes and freeze-dry at -35°C. After drying store the samples at -70°C under
   vacuum. Lyophilization preserves intracellular enzymes and metabolites,
   and allows subsequent manipulation of single cells. Performing this step at -
   35°C is important to prevent changes in metabolites (Passonneau and
   Lowry, 1992). To minimize absorption of water by the samples, all subsequent steps involving manipulation of the freeze-dried cells are done in a
   climate-controlled room at 20°C with relative humidity maintained between
   40–50%.

2. Allow the cells to warm up to 20°C under vacuum, then slowly release the
   vacuum and place the sample on a viewing platform beneath a low power
   binocular microscope. The viewing platform consists of an opalescent plastic top which is illuminated from below by means of a 90° prism. Suspended
   above the top is a radiation source (1–2 µCi of americium) to control static
   electricity. Remove individual myocytes from coverslip fragments by using a
   pin and line up the cells on a sample carrier. Place the carrier containing
   samples in an inverted Petri dish to protect from air currents. The pin used
   to pick up cells is cemented to a small piece of copper wire glued to a pencil-
   shaped dowel rod. Sample carriers are made of a glass strip that has been
   cut from a microscope slide and glued to the end of a small piece of wood.
   When aligned on the sample carrier, it is possible to identify the injected
   myocytes by using a fluorescence microscope.

3. Weigh the myocytes by using a fish pole balance. The balance consists of a
   fine quartz fiber mounted horizontally inside a glass syringe barrel. A glass
   slide is used to cover the opening to protect the balance from air currents.
   The slide is held in place by springs, so that it may be easily opened and
   closed during weighing. An americium source (10–20 µCi) is located inside
   on the bottom of the syringe to dispel static charges. Insert the holder containing the cells into the syringe barrel. Transfer each myocyte to the tip of
   the quartz fiber and measure the deflection. Due to fiber optics, a light
   source positioned appropriately causes the tip of the fiber to glow. A microscope mounted horizontally is focused on the tip and the deflection that occurs with application of the sample is measured by means of micrometer
   ocular in one eyepiece. The deflection is proportional to the weight of the
   sample. The fishpole balance we use for weighing myocytes has an operating range of 3–30 ng. The average weight of a freeze-dried myocyte is approximately 8 ng.

4. Prepare single cell extracts by transferring each myocyte to 0.2 µl of 0.02 N HCl and heating at 80°C for 20 min to destroy enzymes. Then divide the extract into two equal portions (0.1 µl each). The freeze-dried cells disintegrate upon contact with the acidic solution, yielding a homogeneous extract. This step and subsequent incubations involving small volumes are performed under a mixture of 40% hexadecane and 60% USP light mineral oil in Teflon wells. The specific gravity of the mixture is 0.84, so that the aqueous droplets fall to the bottom of the well. The oil well racks used in our laboratory contain 60 wells (3 mm diameter) that have been drilled into a Teflon block (120 mm × 20 mm × 5 mm) leaving a very thin translucent layer at the bottom. The rack is placed on a platform having a black plastic top with a single hole that restricts illumination to a single well. Reagents are added to the oil by using quartz glass constriction pipets, which are capable of delivering volumes <1 nl with an accuracy of 0.5%. Carriers containing the myocytes are suspended above the rack. By means of an eyebrow hair glued to the tip of a pencil-shaped rod, each myocyte is removed from the carrier and pushed through the oil until contact with the aqueous droplet is made.

5. *Conduct the specific reactions to measure ATP.* The reactions involved in measuring ATP are shown in Fig. 1.

   • Add 0.1 µl of the ATP Measurement Reagent **(Solution 1)** to the extract droplet (0.1 µl) and incubate at 25°C for 30 min. Perform incubations in parallel using ATP standards of known concentration.
   • Terminate the reaction by adding 1 µl of 0.05 N NaOH and heating at 80°C for 20 min. In addition to denaturing enzymes, this treatment destroys NADP so that that NADPH formed in the specific reaction can be measured after enzymatic cycling. (Go to step 7)

6. *Conduct the specific reaction to measure 2-deoxyglucose-6-P.* To measure 2-deoxyglucose-6-P accurately, it is necessary to remove glucose-6-P from the extract.

   • Add 0.1 µl of Glucose-6-P Removal Reagent **(Solution 2)** to 0.1 ml extract and incubate at 25°C for 20 min. Heat samples at 80°C for 20 min. This heat-treatment is important as it destroys phosphofructokinase in the removal reagent, thus preventing a back-reaction which would generate glucose-6-P and interfere with the measurement of 2-deoxyglucose-6-P.
   • The reaction involved in measuring 2-deoxyglucose-6-P is shown in Fig. 1.
   • Equilibrate samples at 25°C, add 0.1 µl of 2-Deoxyglucose-6-P Measurement Reagent **(Solution 3)**, and incubate at 25°C for 40 min. Perform incubation in parallel with 2-deoxglucose-6-P standards.
   • Terminate the reaction by adding 0.1 µl of 0.25 N NaOH. Heat samples at 80°C for 20 min to remove excess NADP.

Note:
A relatively high level of glucose-6-phosphate dehydrogenase is required to convert 2-deoxyglucose-6-P to 6-phospho-2-deoxygluconate.
Contaminants present in some commercial preparations of glucose-6-phosphate dehydrogenase result in unacceptably high blank values. Enzyme preparations from Leuconostoc mesenteroides (Calbiochem) have proven to be suitable.

7. *Amplify the NADPH generated in the specific reactions.* The NADP cycle used in measuring ATP and 2-deoxyglucose-6-P is shown in Fig. 2. The Cycling Solution for amplifying NADPH formed in the 2-deoxyglucose-6-P specific reaction is identical except that the concentrations of glutamate dehydrogenase and glucose-6-phosphate dehydrogenase are 300 µg/ml and 50 µg/ml, respectively.

   • Add 1 µl aliquot from the ATP specific reaction (from step 5) to 0.1 ml of the Cycling Reagent **(Solution 4)** and incubate at 38°C for 60 min.

   Note:
   Oil wells are not needed for these reactions, which are conducted in borosilicate glass tubes (10 × 75 mm). NADP standards (0.2 µM) are included to allow determination of cycle number, which is approximately 10000 under these conditions.

   • Terminate the cycling reaction by heating the samples at 95°C for 5 min. The amount of 6-P-gluconate formed during the cycling step is measured in a final reaction.
   • Add 5 µl of the Cycling Reagent **(Solution 4)** to the terminated 2-deoxyglucose-6-P specific reaction mixture.

   Note:
   Due to the relatively small amounts of NADPH generated in the specific reaction, it is necessary to maintain a small volume for these cycling reactions, which are conducted under oil.

   • Incubate samples for 18 h at 25°C. This results in an amplification of approximately 200000 fold.
   • Terminate the cycling reaction by adding 0.5 µl of 1 N NaOH and heating the samples at 80°C for 20 min.

8. *Measure the 6-P-gluconate formed in the cycling reactions.* The indicator reaction used to measure 6-P-gluconate is shown in Fig. 2.

   • Add 1 ml Indicator Reagent **(Solution 5)** to the 100 µl of the terminated cycling mixture containing the 6-P-gluconate generated in the NADP cycle used in the ATP measurements (from step 7).
   • Incubate for 5 min at 23°C, and measure the fluorescence due to NADPH (go to step 9).

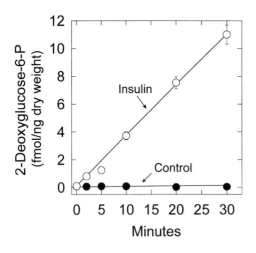

Figure 4 2-Deoxyglucose-6-P accumulation in single cardiac myocytes.

Rat ventricular myocytes were incubated at 37°C without insulin or with 25 milliunits/ml insulin in medium containing 100 mM NaCl, 20 mM KCl, 3 mM MgSO4, 25 mM CaCl2, 1 mM sodium phosphate, 9 mM lactate, 1 mM pyruvate and 10 mM NaHEPES, pH 7.4. After 20 min, the incubations were continued for increasing times in medium supplemented with 100 mM 2-deoxyglucose before the cells were frozen. The results represent 2-deoxyglucose-6-P accumulation expressed in terms of the dry cell weight, and are mean values ± S.E. from at least five cells. Error bars not shown fall within the symbol.

Note:
The amount of 6-P-gluconate formed is determined from a standard curve generated with 6-P-gluconate standards of known concentration.

- Add 5 µl of the terminated Cycling Mixture containing 6-P-gluconate generated in the NADP cycle used in the 2-deoxyglucose-6-P measurements (from step 7) to 1 ml indicator reagent.
- Incubate for 5 min at 23°C, and measure the fluorescence due to NADPH (go to step 9).

9. *Calculate the amount of ATP and 2-deoxyglucose-6-P present in the single cells.* The amounts of ATP and 2-deoxyglucose-6-P are determined from the values obtained with the standards that were included in the respective specific reactions. A time-course of 2-deoxyglucose-6-P accumulation measured in single noninjected myocytes is shown in Fig. 4. In this experiment, the initial rate of accumulation was maintained for at least 30 min and it was markedly increased by insulin due to the stimulatory effect of the hormone on glucose transport. The basal rate of accumulation and the effect of insulin vary among preparations of cells. Damage to the cells by injection may either increase or decrease the rate of accumulation, depending on severity. We have attempted to control for this artifact by discarding cells having ATP concentrations less than 15 mmol/kg dry weight, which is approximately half of the normal ATP concentration. Good microinjection technique is of obvious benefit in limiting cellular damage.

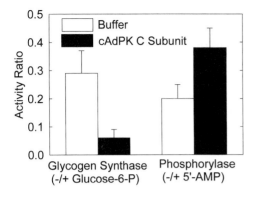

Figure 5   Microinjecting the catalytic subunit of cAMP-dependent protein kinase inactivates glycogen synthase and activates phosphorylase.

Cardiac myocytes were injected with either buffer or with the catalytic subunit (2.5 mg/ml) of beef heart cAMP-dependent protein kinase. Glycogen synthase activities were measured in the absence and presence of glucose-6-P as described in the text. Phosphorylase activities were measured in the absence and presence of 5'-AMP using the assay described by Passoneau and Lowry (1992). The results are expressed as activity ratios and are mean values ± S. E. from 8–10 cells.

3.2   Glycogen synthase activities in microinjected 3T3-L1 cells

Glycogen synthase activities may be readily measured in cardiac myocytes. As shown in Fig. 5, microinjecting the catalytic subunit of cAMP-dependent protein kinase decreased the activity ratio of glycogen synthase and increased the minus AMP/plus AMP activity ratio of glycogen phosphorylase. These effects are consistent with the well established actions of cAMP-dependent protein kinase. Unfortunately, the activation of glycogen synthase produced by insulin in the myocytes is relatively small. 3T3-L1 adipocytes have proven to be a much better system for using microinjection to study the mechanism of action of insulin on glycogen synthase, as insulin produces a marked activation of the enzyme in these cells.

   3T3-L1 fibroblasts are cultured on glass cover slips and differentiated into adipocytes, which are typically used 10–12 days post differentiation. The adipocytes are too small to analyze separately on a routine basis. Therefore, our strategy has been to analyze pools of 20–30 cells. This is accomplished by drawing a small circle (2 mm diameter) on the glass cover slip by using a diamond-tipped pin. A group of 20 to 30 cells near the center of the circle is identified, and every cell in the group is injected with the desired agent in a solution containing rhodamine dextran. The remaining cells in the circle are scraped off of the coverslip by using the microinjection pipet. The injected cells can then be removed and analyzed collectively. It is not practical to weigh the adipocytes, as the cells cannot be readily transferred to and from the fish pole balances. Consequently, the values measured in the cells cannot be normalized based on cellular weight, as is possible with myocytes. This is not a major problem in assessing activation or inactivation of glycogen synthase, as these processes may be monitored by changes in the -/+ glucose-6-P activity ratio, which is independent of absolute units of enzyme activity. For other enzymatic activities and for metabolites, a means of normalization is needed. In theory, it should be possible to express values relative to the number of injected cells, although in practice it is difficult to be sure that every injected cell in a cluster is actually transferred to oil well. Another option is to normalize based on measurements of the activity of an enzyme that does not change in response to the injection protocol. When measured in the presence of glucose-6-P, glycogen synthase is fully active and levels of total synthase activity should be proportional to the number of cells from which the extract was derived. Because of the limited metabolism of 2-deoxyglucose-6-P it is possible to investigate both glycogen synthase and glucose transport in the same microinjected cells (Manchester and Lawrence, 1996).

Protocol 2   Measuring glycogen synthase activities and 2-deoxyglucose-6-P levels in groups of 20–30 3T3-L1 adipocytes

1. Prepare extract from microinjected 3T3-L1 adipocytes.
   - Transfer freeze-dried adipocytes (20–30) by means of a hair point to 0.24 µl of Preincubation Reagent **(Solution 6)** and incubate for 30 min at 25°C.
   - Divide the extract into three 0.08 µl samples for measuring 2-deoxyglucose-6-P and for measuring glycogen synthase activities in the absence and presence of glucose-6-P.

2. Perform specific reaction to measure 2-deoxyglucose-6-P.
   - Add 0.04 µl 0.12 N HCl to 0.08 µl extract and heat at 80°C for 20 min to destroy cellular enzymes.
   - The remaining analytical steps are the same as those described for measuring 2-deoxyglucose-6-P in extracts of myocytes (go to step 6 of Protocol 1).

3. Measure glycogen synthase activity in the absence and presence of glucose-6-P. The reactions involved in assaying glycogen synthase are shown in Fig. 1. For total activity, the Glycogen Synthase Reaction Mixture **(Solution 7)** is supplemented with 20 mM glucose-6-P.
   - Add 0.08 µl of the Glycogen Synthase Reaction Mixture **(Solution 7)** formulated without and with glucose-6-P to 0.08 µl extract samples and incubate at 20°C for 120 min.
   - To terminate the glycogen synthase reactions and to destroy any pyruvate contamination, add 0.08 µl of a solution of 0.09 N NaOH and 6 mM H_2O_2 to each sample. Incubate at 25°C for 30 min.

4. Measure the UDP formed in the glycogen synthase reaction. The reactions involved are shown in Fig. 1.
   - Dissolve the NADH in 50 mM sodium carbonate (pH 10.6) and heat at 95°C for 5 min just prior to use.

- Add 0.08 µl of the UDP Measurement Reagent **(Solution 8)** and incubate samples at 25°C for 30 min.
- Terminate the reaction and remove excess NADH by adding 1 µl 0.1 N HCl and incubating the samples at 25°C for 10 min.

5. Amplify the NAD+ formed. The cycle for amplifying NAD is shown in Fig. 2.

- Add 1 µl of sample containing NAD+ to 0.1 ml of the Cycling Reagent **(Solution 9)** and incubate for 60 min at 25°C.
- Terminate by heating at 95°C for 5 min.

6. Measure the malate formed in the cycle. The reaction used to measure malate is shown in Fig. 2.

- Add 1 ml Malate Indicator Reagent **(Solution 10)** to the terminated cycling reaction, and incubate for 5 min at 25°C.
- Determine the flourescence due to NADH.

7. Determine glycogen synthase activities. Glycogen synthase activities are calculated from the amount of UDP formed, which is determined from the UDP standards included in the specific reactions. The activity ratio is determined by dividing the activity measured in the absence of glucose-6-P from that measured in the presence of glucose-6-P. Typically, incubation with a maximally effective concentration of insulin increases the glycogen synthase activity ratio by approximately three-fold in noninjected cells. Under appropriate conditions injecting buffer alone does not significantly affect the activity ratio, indicating that cells may be injected without disrupting the normal signaling processes that lead to glycogen synthase activation (Manchester and Lawrence, 1996).

References

Manchester J, Kong X, Nerbonne J, Lowry OH, Lawrence JC, Jr. (1994) Glucose transport and phosphorylation in single cardiac myocytes: rate limiting steps in glucose metabolism. *Am. J. Physiol.* **266**: E326-E333.

Manchester J, Kong X, Lowry OH, Lawrence JC, Jr. (1994) Ras signaling in the activation of glucose transport by insulin. *Proc. Nat. Acad. Sci. U. S. A.* **91**: 4644–4648.

Manchester J, Lawrence JC, Jr. (1996) Microanalytical measurements of insulin-stimulated glucose transport in single cells. *Sem. Cell. Dev. Biol.* **7**:279–285.

Mueckler M (1994) Facilitative glucose transporters. *Eur. J. Biochem.* **219**: 713–725.

Passonneau JV, Lowry OH (1992). *Enzymatic Analysis: A Practical Guide.* Human Press, Clifton, NJ.

Skurat AV, Roach PJ (1996) Regulation of glycogen biosynthesis. In: LeRoith D, Olefsky JM, Taylor SI (eds) *Diabetes Mellitus: A Fundamental and Clinical Text.* Lippincott-Raven Publishers, Philadelphia, p 213.

Import of Stably Folded Proteins into Peroxisomes

P.A. Walton

Contents

1 Introduction

Virtually all proteins destined for the peroxisomal matrix and membrane are synthesized on free polysomes in the cytoplasm, and are imported into the peroxisome post-translationally (Lazarow and Fujiki, 1985; Subramani, 1994). Although two peroxisomal proteins, thiolase, and sterol carrier protein 2, undergo proteolytic cleavage after import, most proteins are synthesized at their mature size. This differs from the mechanism of post-translational import of proteins into the chloroplast, and mitochondria, where the targeting signal resides as an amino-terminal leader sequence that is cleaved following import. Import of proteins into the matrix of the peroxisome is dependent upon the presence of a peroxisomal targeting signal in the amino acid sequence of the

Methods and Tools in Biosciences and Medicine
Microinjection, ed. by J. C. Lacal et al.
© 1999 Birkhäuser Verlag Basel/Switzerland

newly synthesized protein. Two forms of this peroxisomal targeting signal have been characterized: a C-terminal tripeptide with the sequence serine-lysine-leucine or a conservative variant thereof which has been found in many peroxisomal proteins (Gould et al., 1989), or an amino-terminal leader sequence such as that found in peroxisomal thiolase (Swinkels et al., 1991).

Further elucidation of the mechanisms of peroxisomal protein import will be facilitated by the use of *in vitro* systems. Unfortunately, the fragility of purified peroxisomes, the lack of an easily identifiable hallmark for import (such as signal cleavage used to study translocation in the ER, mitochondria and chloroplasts), and the inefficient import observed to date have combined to prevent the description of all but the most elementary aspects of peroxisomal protein import. As a first step towards the development of a reliable *in vivo* import system that offers many of the advantages of *in vitro* import assays, we have established a microinjection system for the study of peroxisomal protein import. In the present report, I have described the features of peroxisomal protein transport in mammalian cells observed by following the import of natural and chemically-modified protein substrates into peroxisomes following microinjection.

2 Materials

Chemicals and cell culture

Anti-catalase antibodies	Calbiochem, La Jolla, CA, USA
Anti-luciferase antibodies	Promega, Madison, WI, USA
Dehydroluciferin	Sigma, St. Louis, MO, USA
DSP	Pierce Chemical Company, Rockford, IL, USA
Fetal calf serum	GIBCO-BRL, Gaithersburg, MD, USA
Firefly luciferase	Sigma, St. Louis, MO, USA
Fluorescently-labeled secondary antibodies	Jackson Immuno Research Laboratories, West Grove, PA, USA
Fluorescently-labeled streptavidin	Amersham, Arlington Heights, IL, USA
Human fibroblast cells (Hs68)	American Type Culture Collection, Rockville, MD, USA
Human serum albumin	Sigma, St. Louis, MO, USA
Luciferin	Sigma, St. Louis, MO, USA
NHS-biotin	Pierce Chemical Company, Rockford, IL, USA
Rabbit polyclonal antibodies against human serum albumin	ICN, Costa Mesa, CA, USA
SPDP	Pierce Chemical Company, Rockford, IL, USA
Sulfo-MBS	Pierce Chemical Company, Rockford, IL, USA
Synthetic peptides	Agouron Institute or Multiple Peptide Systems, La Jolla, CA, USA

Other reagents were purchased from the standard sources.

Equipment

• Centricon 30 filters, Amicon, Danvars, MA, USA

3   Methods

3.1   Cross-linking of biotin and peptides bearing the peroxisomal targeting signal (PTS) to human serum albumin

Protocol 1   Cross-linking of biotin and peptides to human serum albumin

1. Human serum albumin, at a concentration of 10 mg/ml in 50 mM $NaHCO_3$ (pH 8.5), was incubated with 7.6 mM sulfo-MBS and NHS-biotin for 1 h at room temperature.

2. Excess reagents were neutralized by the addition of Tris to 40 mM and separated from the MBS-linked proteins by Centricon filtration.

3. The modified albumin was subsequently incubated with the synthetic peptide NH_2-CRYHLKPLQSKL-COOH, overnight at 4°C.

4. The cross-linked products were separated from the unreacted peptide, and the buffer changed to PBS by Centricon filtration.

5. Approximately 5–15 such peptides were attached to each HSA molecule as judged by an increase in apparent molecular weight using SDS polyacrylamide gel electrophoresis.

   Note:
   Due to the nature of the cross-linking chemistry, it is unlikely that a peptide could have attached to the C-terminus of the protein. Therefore the peptides must have been attached as side-chains to lysine residues on the HSA molecule and none could have been co-linear with the protein.

The resulting hybrid molecule was referred to as bHSA-SKL. A similar hybrid molecule, with a reducible disulfide between the protein and the PTS (bHSA-s-s-SKL) was prepared as described above, except that the sulfo-MBS was replaced with the disulfide-containing crosslinker SPDP.

3.2   Cell culture

Hs68 cells were grown in DMEM supplemented with 10% fetal calf serum. For microinjection, cells were plated on glass coverslips, which had been acid-washed (2 parts concentrated H_2SO_4 : 1 part concentrated H_3NO_2), extensively rinsed in 5 mM EDTA and deionized water, stored in absolute ethanol, and flame-sterilized before use.

3.3   Microinjection and immunofluorescence

Cells were microinjected using glass capillary needles, made using a Kopf vertical pipet puller (model 720).

Protocol 2   Microinjection

1. Luciferase was microinjected at a concentration of 0.2 mg/ml in a buffer of 20 mM KPO_4 (pH 7.4), 100 mM KCl, and 40 mM potassium citrate. With an average injection volume of 5×10^{-14} liters, a molecular weight of 62 kdal, and a concentration of 0.2 mg/ml, approximately 1×10^5 molecules of luciferase were injected per cell. In addition, injections included approximately 2×10^5 molecules of mouse IgG.

2. Human serum albumin, containing the cross-linked peroxisomal targeting signal (bHSA-SKL), or containing the disulfide-linked cross-linked peroxisomal targeting signal (bHSA-s-s-SKL), was microinjected at a concentration of 0.5 mg/ml; approximately 2×10^5 molecules of albumin were injected per cell.

3. To facilitate identification of microinjected cells, mouse IgG (non-specific) was co-injected at a concentration of 1 mg/ml.

Following microinjection, the cells were returned to the incubator and routinely incubated for 16 h at 37°C. Following the incubation the cells were fixed, permeabilized, and immunostained as follows.

Protocol 3   Immunofluorescence

1. Cells on coverslips were washed in PBS.

2. Cells were fixed in 3.7% formaldehyde in PBS for 10 min.

3. Cells were permeabilized with 0.1% Triton X-100 in PBS for 5 min followed by washing with 0.01% Tween 20 in PBS in this and subsequent washes.

4. A mixture of primary antibodies (1:100 dilution) was applied to the coverslips in a humidified chamber at 1:100 dilution and incubated for 30 min at room temperature.

5. The cells were washed and a mixture of secondary reagents consisting of FITC-labelled goat anti-rabbit and Texas Red-labelled streptavidin (both at 1:100 dilution) was applied to the coverslips and incubated for 30 min.

6. Cells were washed extensively, rinsed in H_2O and coverslips mounted on microscope slides for observation.

Fluorescence microscopy was performed with a Zeiss Axiophot Photomicroscope or a Zeiss LSM 420 confocal microscope using a 63× (1.3NA) lens. Photographs of the fluorescent images were recorded using Kodak T-Max 400 film, pushed and developed one stop as per the manufacturers instructions. Identical x,y,z-axis confocal sections were made of the Texas Red and FITC fluorescence, using the 543nm (HeNe) and 480nm (ArKr) lasers. False-colour overlays of the two sections were prepared at the time of image collection. Digitized images were stored and printed without further modification using a Kodak XLS 8300 digital printer.

3.4   Controls

As an important control, we sought to confirm that transport of bHSA-SKL was into the lumen of the peroxisomes, and not merely aggregation of the bHSA-SKL molecules on the surface of the vesicles. In order to ascertain whether the bHSA-SKL was transported into membrane-enclosed vesicles, cells microinjected with bHSA-SKL, and incubated 16 h at 37°C were fixed, permeabilized with either digitonin or digitonin plus Triton X-100, and stained as before, except that no detergents were used in either the buffers ar washes. Digitonin permeabilization of mammalian cells at a concentration of 25 µg/ml for 10 min has been demonstrated to allow access to the cytoplasmic compartment while retaining the integrity of the peroxisomal membranes (Wolvetang et al., 1990). After an incubation of 16 h at 37°C following injection of the bHSA-SKL conjugate, cells permeabilized with digitonin demonstrated staining of the cytoplasmic mouse IgG but the bHSA-SKL was not visible. Permeabilization of identically injected and incubated cells with digitonin plus Triton X-100 revealed spherical bHSA-SKL containing structures in injected cells. As an internal control, cells permeabilized with either digitonin or digitonin plus Triton X-100 were stained for endogenous catalase. In cells permeabilized with digitonin the antibodies failed to reveal the peroxisomally located catalase, but showed the normally observed punctate pattern when permeabilized with digitonin plus Triton X-100 prior to staining. This demonstrates that the bHSA-SKL was sequestered, like catalase, within the peroxisomal matrix.

3.5   Preparation of colloidal gold and electron microscopy

Colloidal gold particles were prepared by the reduction of gold chloride (Slot and Geuze, 1985), and were coated with 1.5 mg bHSA-SKL and subsequently with 100 mg HSA as an additional stabilizer. Colloidal gold/protein complexes were isolated and separated from unbound proteins by centrifugation, and the buffer changed to injection buffer by Centricon filtration. For electron microscopic examination, Hs68 cells were grown in 35 mm plastic tissue culture dishes. To facilitate the identification of microinjected cells, circles of approximately 2 mm were drawn on the bottom of the dishes and all cells within the circles were microinjected. Following microinjection of the gold/bHSA-SKL complex, the cells were incubated for 16 h at 37°C. Cells were fixed with 4% paraformaldehyde and 0.1% glutaraldehyde, post-fixed with 1% osmium tetroxide, dehydrated with graded alcohols, and embedded in Epon. Blocks containing the microinjected cells were sawn out of the Epon pucks, sectioned and mounted on grids, and stained with uranyl acetate and lead citrate prior to observation. For immunoelectron microscopy the osmium tetroxide step was omitted. These sections were mounted on gold grids, etched with 10% H_2O_2, and immunostained with antibodies directed against catalase (1:5 dilution in PBS plus 2% BSA, 2% casein, 0.5% ovalbumin). The secondary reagents were 10 nm gold-labeled goat anti-rabbit antibodies (Sigma) (1:20 dilution). Electron microscopy was performed with a Phillips 400T electron microscope at 80 KeV.

4   Procedures and Applications

4.1   Microinjection of purified peroxisomal proteins into mammalian fibroblasts: import of these proteins into the peroxisomal matrix.

Luciferase
Luciferase was microinjected into Hs68 cells, incubated overnight under normal growth conditions and processed for double-label indirect immunofluorescence using anti-luciferase and anti-catalase (an endogenous peroxisomal matrix protein) antibodies. As shown in Figure 1a, the microinjected luciferase was present in vesicular structures very similar in appearance to peroxisomes. To confirm that these vesicles were indeed peroxisomes, costaining for luciferase and catalase resulted in identical punctate patterns (Figs 1a and 1b, respectively).

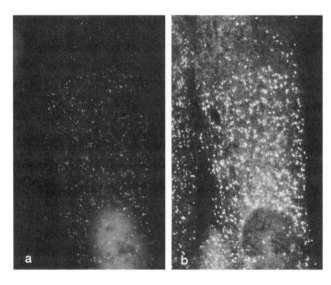

Figure 1 Colocalization of microinjected luciferase (a) with endogenous catalase (b).

Following injection of approximately 10^5 molecules of luciferase and incubation for 16 h at 37°C, the cells were processed for double indirect immunofluorescence. This employed guinea pig anti-luciferase and rabbit anti-catalase antibodies in the first step, and FITC-conjugated anti-guinea pig IgG and rhodamine-conjugated anti-rabbit IgG antibodies in the second step.

The transport of luciferase was time dependent. The protein began to appear within peroxisomes after 2 h at 37°C. The number of such vesicles increased through 4 and 8 h and appeared to reach a plateau by 18 h. Interestingly, the number of vesicles, but not their final size, increased with the amount of luciferase injected. The process appeared to be saturable because microinjection of high concentrations of luciferase (an order of magnitude greater than that used in Fig. 1) resulted in considerable cytoplasmic staining following overnight incubations.

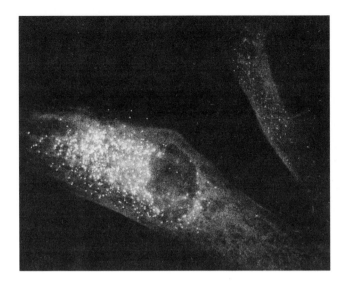

Figure 2 Import of luciferase-dehydroluciferin-AMP complex.

Hs68 cells were microinjected with the dead-end enzyme-substrate complex, incubated for 16 h at 37°C, and processed for double indirect immunofluorescence. Figure shows the intracellular location of luciferase complex in microinjected cells.

To confirm the requirement for the peroxisomal targeting signal in peroxisomal transport, synthetic peptides bearing the peroxisomal targeting signal SKL (NH₂-CRYHLKPLQSKL-COOH) were coinjected with luciferase at various concentrations and the cells were incubated for 16 h at 37°C. The results indicated that this peptide did not appreciably diminish transport at a 10-fold and 20-fold molar excess. At a 50-fold or 100-fold molar excess the transport of luciferase into peroxisomes was abolished. Coinjection of other peptides bearing the related peroxisomal targeting signals AKL, and SRL at a 50-fold molar excess also resulted in the total inhibition in luciferase transport. A control peptide bearing the first nine amino acids of the SKL inhibitory peptide had no effect on transport at a 100-fold molar excess.

Preincubation of luciferase with dehydroluciferin and ATP resulted in the formation of a dead-end enzyme-substrate complex. This luciferase-dehydroluciferin-AMP complex is extremely stable, with a K_d of 5×10^{-10}. Such complexes have been used to demonstrate that proteins destined for import into the mitchondria must be unfolded in order to be imported efficiently (Eilers and Schatz, 1986). However, in peroxisomes, proteins with stably-associated substrates are imported in a manner that is indistinguishable from normal proteins (Fig. 2).

Import of hybrid peroxisomal proteins
In order to determine whether artificial substrates could be imported into peroxisomes, we created a protein-peptide conjugate in which dodecameric peptides ending in SKL-COOH were cross-linked to lysines in human serum albumin (HSA). The bHSA-SKL conjugate was microinjected into Hs68 cells, and after 16 h the bHSA-SKL conjugate was observed in the peroxisomal compartment (Fig. 3). Import was time dependent. Microinjection of unconjugated HSA

Figure 3 Import of bHSA-SKL into peroxisomes.

Hs68 cells were microinjected with approximately 2×10^5 molecules of bHSA-SKL and incubated for 16 h at 37°C. Cells were subsequently fixed, and immunostained for bHSA-SKL and endogenous catalase. Figure shows the intracellular location bHSA-SKL (a) and catalase (b) in a microinjected cell.

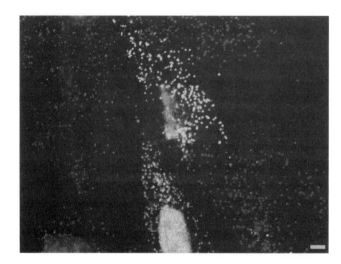

Figure 4 Peroxisomal import of microinjected bHSA-s-s-SKL.

Hs68 cells were microinjected with bHSA-s-s-SKL, incubated for 16 h at 37°C, and immunostained for bHSA-s-s-SKL and endogenous catalase. Staining consisted of rabbit anti-catalase as the primary reagent and FITC-conjugated donkey anti-rabbit antibodies and streptavidin-conjugated Texas Red as secondary reagents. Identical confocal microscopic sections were scanned for both fluorochromes. Figure shows the false-color overlay indicating vesicles that contain catalase (green), bHSA-s-s-SKL (red), or both (yellow). Import of bHSA-s-s-SKL was into endogenous peroxisomes as confirmed by costaining for catalase in the two microinjected cells; peripheral cells were uninjected and stain only for catalase. Bar, 5mm.

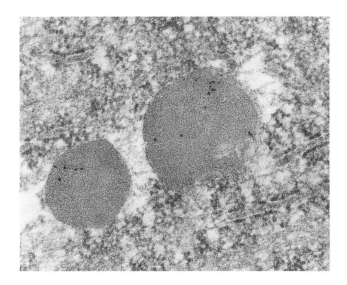

Figure 5 Peroxisomal import of gold particles bearing the SKL peroxisomal targeting signal.

Colloidal gold particles (4–9 nm diameter), coated with bHSA-SKL were prepared as described in Materials and Methods. Following microinjection into the cytosol of Hs68 cells and incubation for 16 h at 37°C, cells were fixed and prepared for electron microscopy. Figure shows the intracellular location of the microinjected gold particles. Microinjected gold/bHSA-SKL complexes were detected within electron-dense vesicles, confirmed to be peroxisomes by the presence of immunologically-detectable catalase.

into fibroblasts did not result in the import of this protein into the peroxisomal compartment after an incubation of 16 h.

HSA contains 17 disulfide bonds. Although the reductive environment of the cytosol is predicted to prevent disulfide formation (Hwang et al., 1992), it is not known whether a protein that contains disulfide-bonded cysteines could be reduced. As a test of the stability of disulfide bonds, bHSA-s-s-SKL prepared with a disulfide bond in the crosslinker was microinjected into Hs68 cells and incubated for 16 h at 37°C. This molecule was imported into peroxisomes (Fig. 4), indicating that the disulfide bond was stable in the cytosol. With stable disulfide bonds, the microinjected albumin molecules are expected to have retained much of their tertiary structure during the import assays.

Import of colloidal gold particles bearing the peroxisomal targeting signal into the matrix of the mammalian peroxisome

In order to determine if the peroxisomal import machinery could import a large, nondeformable complex, colloidal gold particles coated with bHSA-SKL were microinjected into Hs68 cells. After the standard incubation time of 16 h, the microinjected cells were processed for electron microscopy. Electron micrographs clearly show colloidal gold particles within the limits of single-membrane bound structures (Fig. 5). These structures were confirmed to be peroxisomes by the presence of catalase. The specificity of this import was confirmed by the observation that few, if any, gold particles were found in association with other subcellular compartments. As an additional control, coinjection of a synthetic peptide bearing a peroxisomal targeting signal which has previously been shown to inhibit peroxisomal import (Walton et al., 1992a), resulted in a reduction by approximately two-thirds in the number of peroxisomal profiles containing gold particles and a reduction in the average number of gold particles observed per peroxisomal profile by approximately 85%. These experiments demonstrate that the gold particles (and hence the bHSA-SKL) gain entry into the peroxisomal matrix in a PTS1-dependent manner. Electron microscopic examination of the imported gold particles indicated that their diameters ranged from 4–9 nm. Our results confirm that multimeric complexes can be imported into peroxisomes, even if one of the constituents is a metal particle. Furthermore, the dimensions of a particle that can traverse the peroxisomal membrane is at least 9 nm, the largest gold particle observed within the peroxisomal matrix.

5   Remarks and Conclusions

Microinjection of purified proteins into mammalian cells has been an effective tool in the study of the cell (Feramisco and Welch, 1986). This chapter describes the transport of microinjected proteins bearing the peroxisomal targeting signal into peroxisomes of mammalian cells. The presence of the peroxisomal targeting signal SKL-COOH, either as a constituent of the purified firefly luciferase or as part of synthetic peptides crosslinked to human serum albumin was necessary and sufficient to direct this translocation *in vivo*. The re-translocation of luciferase unequivocally demonstrates that no irreversible alteration of the tripeptide PTS occurs upon import into peroxisomes and is in fact the first demonstration that a protein can undergo multiple translocations across the same type of membrane. This type of protein translocation would not be possible for the majority of proteins destined for the ER, mitochondrion or chloroplast since these proteins lose their targeting signals by proteolytic cleavage upon import. However, re-transport of a nuclear protein through the nuclear pore complex has been demonstrated for the catalytic subunit of cAMP-dependent protein kinase following microinjection (Meinkoth et al., 1990). This must obviously occur often as a method of transcriptional control and in the normal course of events since the nucleus must reassemble after each cell division.

Given that the microinjection system can be used to study aspects of peroxisomal protein import, we have exploited this system to answer some of the many questions regarding this process. My lab has used this system to demonstrate the lack of peroxisomal import in cells derived from patients with Zellweger syndrome (Walton et al., 1992a), import of a yeast peroxisomal protein in mammalian cells (Walton et al., 1992b), transport-competence of virtually all peroxisomes (Hill and Walton, 1995), involvement of hsp70 molecules (Walton et al., 1994), and the import of stably folded proteins highlighted above

(Walton et al., 1995). In addition, the microinjection system provides the means to answer further questions concerning peroxisomal protein import and biogenesis; questions limited only by the imagination of the investigator.

Acknowledgements

I would like to thank Dr. Suresh Subramani, Dr. Bill Welch, and Dr. Jim Feramisco, for their support and encouragement. This work has been supported by grants from the Medical Research Council of Canada.

References

Feramisco JR, Welch WJ (1986) Modulation of cellular activities via microinjection into living cells. In: Microinjection and Organelle Transplantation Techniques (Celis, J.E., Graessmann, A., and Loyter, A. eds.) Academic Press (London) pp. 39–58.

Gould SJ, Keller G-A, Hosken N et al. (1989) A conserved tripeptide sorts proteins to peroxisomes. J. Cell Biol. 108: 1657–1664.

Hill PE, Walton PA (1995) Import of microinjected proteins bearing the SKL peroxisomal targeting sequence into the peroxisomes of a human fibroblast cell line: evidence that virtually all peroxisomes are import-competent. J. Cell Sci. 108: 1469–1476.

Hwang C, Sinskey AJ, Lodish HF (1992) Oxidized redox state of glutathione in the endoplasmic reticulum. Science 257: 1496–1502.

Lazarow PB, Fujiki Y (1985) Biogenesis of peroxisomes. Ann Rev Cell Biol 1: 489–530.

Meinkoth JL, Ji Y, Taylor SS, Feramisco JR (1990) Dynamics of the cyclic AMP-dependent protein kinase distribution in living cells. Proc. Nat. Acad. Sci. USA 87: 9595–9599.

Slot JW, Geuze HJ (1985) A new method of preparing gold probes for multiple-labeling cytochemistry. Eur. J. Cell Biol. 38: 87–93.

Subramani S (1993) Protein import into peroxisomes and the biogenesis of the organelle. Annu. Rev. Cell Biol. 9: 445–478.

Swinkels BW, Gould SJ, Bodnar AG et al. (1991) A novel, cleavable peroxisomal targeting signal at the amino-terminus of the rat 3-ketoacyl-CoA thiolase. EMBO J. 10: 3255–3262.

Walton PA, Gould SJ, Feramisco JR et al. (1992a) Transport of microinjected proteins into peroxisomes of mammalian cells: Inability of Zellweger cell lines to import proteins with the SKL tripeptide peroxisomal targeting signal. Mol. Cell. Biol. 12: 531–541.

Walton PA, Gould SJ, Rachubinski RA et al. (1992b) Transport of microinjected alcohol oxidase from Pichia pastoris into vesicles in mammalian cells. Involvement of the peroxisomal targeting signal. J. Cell Biol. 118: 499–508.

Walton PA, Wendland M, Subramani S et al. (1994) Involvement of 70 kDa heat-shock proteins in peroxisomal import. J. Cell Biol. 125: 1037–1046.

Walton PA, Hill PE, Subramani S (1995) Import of stably-folded proteins into peroxisomes. Molec. Biol. Cell 6: 675–683.

Wolvetang EJ, Tager JM, Wanders RJA (1990) Latency of the peroxisomal enzyme acyl-CoA: dihydroxyacetonephosphate acyltransferase in digitonin-permeabilized fibroblasts: the effect of ATP and ATPase inhibitors. Biochem. Biophys. Res. Commun. 170: 1135–1143.

6 Study of the Function of Microtubule Proteins

R. Armas-Portela and J. Avila

Contents

1 Introduction

Microtubules are one of the main cytoskeletal structures present in eukaryotic cells. They are involved in many cellular functions such as chromosome segregation, intracellular transport, in the establishment of cell morphology and differentiation (Avila, 1990). They are mainly isolated from brain tissue where they are present in large amount. Brain microtubules are composed of tubulin, the major microtubule component, and a lower proportion of other proteins known as microtubule-associated proteins (MAPs) (Matus, 1988). MAPs have been classified into three groups; the first group comprises MAP1 proteins, the second one includes MAP2 proteins and the third one, tau proteins. Furthermore, non-neural cells have other microtubule-associated proteins of which

Methods and Tools in Biosciences and Medicine
Microinjection, ed. by J. C. Lacal et al.
© 1999 Birkhäuser Verlag Basel/Switzerland

the family of MAP-4 proteins is the best characterized (Bulinski and Borisy, 1980; Aizawa et al., 1990).

Microtubules are intrinsically dynamic polymers that show spontaneous and rapid subunit exchange with unassembled tubulin at steady state and dynamic instability (Mitchison and Kirschner, 1984). *In vitro* MAPs regulate microtubule dynamic properties (Pryer, 1992) and several studies reveal an involvement of MAPs in microtubule stabilization (Hotani and Horio, 1988). This mechanism may provide an explanation of how microtubules contribute to cellular morphology and differentiation, and suggest that stabilization of microtubules constitutes the basis of the formation of axon and dendrites, responsible for the functional polarity of nerve cells.

Microinjection has been a valuable method to evaluate the effect of several factors on the polymerization or stabilization of microtubules. Microinjection of soluble tubulin unexpectedly resulted in a suppression of tubulin synthesis "*de novo*" (Cleveland et al., 1983). On the other hand, microinjection of tau protein into fibroblasts increased tubulin polymerization and stabilised microtubules against depolymerization (Drubin and Kirschner, 1986). Furthermore, a rapid turnover of MAP2 occurs in the axon of neurons (Okabe and Hirokawa, 1989), although in fibroblast cells the effect of microinjected MAP2 seems to be depend on the MAP phosphorylation state (Brugg and Matus, 1991). Since there are different MAPs that contain similar tubulin-binding domains (Chapin and Bulinski, 1991) microinjection of synthetic peptides corresponding to these domains and antibodies against them may serve as an approach to assess the significance of MAP-tubulin interactions on microtubule organization. In this chapter we present the protocols for microinjection of synthetic peptides of microtubule-associated proteins responsible for tubulin binding and the microinjection of specific antibodies against these domains. In order to facilitate the incorporation of synthetic peptides and the interaction of MAPs with the injected antibodies, microinjection was performed in cells treated with nocodazol to depolymerize the microtubules. After treatment with this depolymerizing drug, cells are able to recover the normal microtubule pattern, but microinjected cells showed microtubule rearrangements and the appearance of aberrant structures. Similar approaches should be possible using other particular domains of MAPs to further our understanding of the regulation of microtubule dynamics.

2   Materials

Chemicals and cell culture

Ammonium sulphate	Sigma
Bovine serum albumin (BSA)	Sigma
Dulbecco Modified Eagle's Medium (DMEM)	Gibco, BRL
Fetal calf serum	Gibco, BRL
FITC-conjugated anti-mouse IgGs	Jackson Immunol. Res., Inc.
FITC-dextran, 150000 MW	Sigma
Fluoromount-G	Southern Biotechnology Associates, Inc.
Hepes (N-[2-Hydroxyethyl]piperazine-N'-[2-ethanesulfonic acid])	Sigma
Mouse anti-tubulin monoclonal antibody	Amersham
Nocodazol	Sigma
poly-D-lysine	Sigma
Texas-Red conjugated goat anti-mouse IgGs	Amersham
Texas-Red conjugated goat anti-rabbit IgGs	Jackson Immunol. Res., Inc.

Equipment

- Cellocate coverslips
- Fluorescence microscope
- HPLC apparatus
- Microconcentrators (Centricon 30, Amicon)
- Microfuge
- Microinjector
- Microloaders (Eppendorf, Cat 542 956 003)
- Sterile micropipets (Eppendorf, Cat 542 952 008)

Solutions

- **Injection buffer**
    20 mM 2.N-morpholinoethane-sulfonic acid (MES) pH 7.5
- **Phosphate buffered saline (PBS)**
    130 mM NaCl
    7 mM Na2HPO4
    3 mM NaH2PO4 pH 7.4

3   Methods

3.1   Cell culture and preparation of coverslips

Cells are seeded on coverslips in cell culture plates. We used Vero cells grown at low density for 12 h in DMEM supplemented with 5% fetal calf serum (FCS) at 37°C in a humidified 5% CO2 atmosphere. 1 µM nocodazol 3 h before the injection was added when microtubule depolymerization was required.

a. Coat coverslips with a sterile poly-D-lys solution (10 µg/ml) and incubated for 37°C

b. Eliminate poly-D-lys excess and wash coverslips with sterile water before seeding the cells.

To facilitate the location of injected cells for their analysis under the microscope, a small area can be selected by carbon shadowing 200 mesh electron microscopy grids over the coverslips followed by sterilisation. Alternatively commercial sterile cellocate coverslips can be used (Eppendorf, Cat. 5245 962.004)

3.2   Microinjection of synthetic peptides

Selection and Purification of Synthetic Peptides
Synthetic peptides corresponding to particular domains of microtubule-associated proteins can be designed. This is achieved by studying protein sequences found in universal databases and selecting the domain of interest. We select a peptide (tau 183–204: KVTSKCGSLGNIHHKPGGG) corresponding to the second repeat of the tubulin binding domain of tau protein. This peptide was purified by reverse phase HPLC on a Nova-Pack C18 column using a Waters apparatus and lyophilised. However, similar approaches should apply to other synthetic peptides corresponding to other domains.

Protocol 1   Preparation of synthetic peptide for microinjection

1. Dissolve purified peptide in bidistilled water as a stock solution (3 mg/ml)

2. Filter the solution with a 0.22 µm Millipore and store at −20°C

3. For injection prepare the peptides at different concentrations (1–2 mg/ml) in injection buffer in the presence of 5% of the fluorescence marker FITC-dextran, MW 150000 (Sigma)

4. To avoid clogging the needle, centrifuge the injection sample at 14000 g for 15 min at 4°C immediately before use.

3.3 Microinjection of antibodies

Preparation and purification of antibodies
Antibodies against the selected synthetic peptides can be raised by rabbit immunization (see Harlow and Lane, 1988). IgGs are precipitated and concentrated according the following steps.

Protocol 2 IgG precipitation and concentration

1. Add ammonium sulphate (pH 7.4) to the serum until 50%, at 4°C

2. Centrifuge 30000 rpm 20 min, 4°C

3. Resuspend in 0.5 volumes of PBS. Aliquot and store at -20°C

4. Add ammonium sulphate (pH 7.4) to the serum until 50%, at 4°C

5. Centrifuge 30000 rpm 20 min, 4°C

6. Resuspend in 0.5 volumes of PBS. Aliquot and store at -20°C

Preparation of Sera for Microinjection

Protocol 3 Preparation of sera for microinjection

1. For injection, antibodies are dialysed overnight against injection buffer

2. Concentration with microconcentrator (Centricon 30, Amicon) is recommended

3. Aliquot and store at 4°C until use.

Microinjection
Synthetic peptides or antibodies in injection buffer are loaded in sterile micropipets using a microloader. Microinjection was performed using an inverted microscope (Axiovert 35, Zeiss, Germany) equipped with an automated injection system (AIS) (Zeiss, Germany) and an Eppendorf microinjector (Model 5242). To avoid pH changes during the microinjection, the medium was buffered with 25 mM Hepes (pH 7.4). Usually 90–100 cells were injected in each experiment. Control cells were injected with FITC-dextran (0.5 %) in injection buffer. The volume of injection was about 5% of the total cellular volume and established after a preliminary set of automated injection times and pressures. In those experiments where microtubule depolymerization was required to favor the interaction of the peptide with microtubules, cells were treated for 3h with nocodazol (1 μM) which was also present during microinjection. After the injection, cells were washed and incubated in fresh medium for different times before fixation for immunofluorescence.

Protocol 4 Microinjection

1. Load the micropipet with 0.5–1 µl of injection sample avoiding bubbles

2. Insert the micropipet into the holder of the microinjector

3. Attach the holder to the micromanipulator and rapidly introduce the micropipet tip in the medium of the culture dish. This step is very important to avoid clogging the needle.

3.4 Immunofluorescence analysis

A different protocol for immunofluorescence will be applied depending on whether the microinjection of synthetic peptides or the corresponding antibodies will be analyzed.

Peptide microinjection analysis

Protocol 5 Peptide microinjection analysis

1. Wash coverslips with PBS

2. Fix in methanol -20°C for 15 min

3. Wash fixative with PBS

4. Block non-specific groups with 1% BSA in PBS for 30 min at room temperature

5. Incubate with a mouse monoclonal anti-tubulin antibody prepared in 1% BSA in PBS according to the appropiate dilution, 1h at room temperature

6. Wash 3×5 min in PBS

7. Incubate with a Texas Red anti-mouse IgGs diluted in 1% BSA in PBS, 30 min

8. Wash 3×5 min in PBS and mount with Fluoromount G

Antibodies microinjection analysis

Protocol 6 Antibodies microinjection analysis

1–6. Process as the Protocol 5

7. Incubate the coverslips with a mixture of Texas Red conjugated anti-rabbit and FITC-conjugated anti-mouse antibodies at a suitable dilution in 1%BSA in PBS

8. Wash 3×5 min with PBS

9. Mount the coverslips over microscope slides using Fluoromount G

4 Remarks

The previous protocols were used to study if the injection of the region containing a tubulin binding motif of tau could be enough to stabilize microtubules (Drubin and Kirschner, 1986) or to induce microtubule rearrangements, as it has been indicated for the overexpression of MAP2, which shares this tubulin binding motif (Edson et al., 1993). Figure 1 shows that microinjection of this peptide of tau in Vero cells previously treated with nocodazol and allowed to recover after the injection, resulted in the appearance of microtubule bundles not linked to centrosomes. Microtubules were bent around the periphery of the cell. This bending may result from rigidity of microtubule bundles and appear as a consequence of microtubules longer than the cell diameter. In this situation microtubules will encounter resistance at the cell surface and thus bend. Likewise, the appearance of cytoplasmic extensions seems to require rearrangements of the cortical cytoskeleton (Edson et al., 1993). However, it is noteworthy that the presence of a unique tubulin binding motif of tau protein could be enough for the formation of microtubule bundles. Therefore, the lack of the amino and carboxy terminal regions could favor the induction of bundles, as previously suggested by Brandt and Lee (1993). In this way it has been suggested (Hirokawa et al., 1988) that the extreme ends of tau may protude from the microtubule forming a projection that may prevent a closed interaction of microtubules.

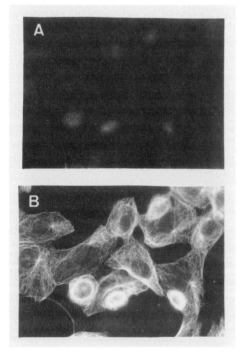

Figure 1 Effect of microinjection in Vero cells of a tau synthetic peptide corresponding to the second repeat of tubulin binding domain in tau.

The cells were previously treated with 1 μM nocodazol to depolymerize the preexisting microtubules and microinjection was performed in its presence. After 20 min of recovery postmicroinjection, the coverslips were processed for immunofluorescence. A) Fluorescein fluorescence showing FITC-dextran present in microinjected cells. B) Texas Red anti-tubulin staining of cells recovered after nocodazol removal. Microinjected cells show microtubule bundles bent at the cell periphery, whereas a normal microtubule pattern radiating from the centrosome is observed for non injected cells.

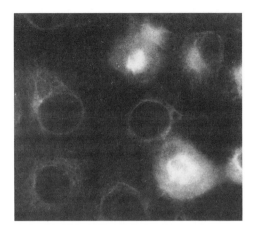

Figure 2 Effect of anti-369 antibody microinjection on microtubules of Vero cells.

Texas Red fluorescence for anti-369 microinjected cells and fluorescein antitubulin staining were photographed on the same frame.

In an alternative approach, microinjection of antibodies raised against specific domains may be used as a way for searching cell proteins having common functional motifs. In Vero cells there are proteins recognized by this antibody (Vial et al., 1995). A possible endogenous protein related to tau protein is MAP4 which contains tubulin binding domains with a sequence related to that of tau and MAP2 (Chapin and Bulinski, 1991)). Figure 2 shows that unexpectedly, a great number of microinjected cells showed aberrant spindle structures. The occurrence of such abnormal phenotypes in mouse or human cells exposed to antisense oligonucleotides of MAP4 has also been described (Olson and Olmsted, 1993). Nevertheless, no observable cell phenotype was observed using antibodies against the MAP4 carboxy- terminus for microinjection (Wang et al., 1996). A possible explanation is that the presence of extra regions on both sides of a particular motif could regulate the interaction of certain important domains with microtubules (Goode et al., 1997).

References

Aizawa H, Emori Y, Murofushi H, Kawasaki H, Sakai H. and Suzuki K (1990) Molecular cloning of a ubiquitously distributed microtubule-associated protein with Mr 190000. J. Biol.Chem. 266: 13849–13855.

Avila J (1990) Microtubule dynamics. FASEB J. 4: 3284–3290.

Brandt R, Lee G (1993) Functional organization of microtubule associated protein tau. J. Biol. Chem. 268: 3414–3419.

Brugg B, Matus A (1991) Phosphorylation determines the binding of microtubule-associated protein 2 (MAP2) to microtubules in living cells. J. Cell Biol. 114: 735–743.

Bulinski JC, Borisy GG (1980) Widespread distribution of a 210000 mol wt microtubule-associated protein in cells and tissues of primate. J. Cell Biol. 87: 802–808.

Cleveland DW, Pittenger MF, Feramisco J (1983) Elevation of tubulin levels by microinjection suppress new tubulin synthesis. Nature 305: 738–740.

Drubin DG, Kirschner MW (1986) Tau protein function in living cells. J.Cell Biol. 103: 2739–2746.

Edson K, Weisshaar B, Matus A (1993) Actin depolymerisation induces process formation in MAP2 transfected non neural cells. Development 117: 689–700.

Goode BL, Denis PE, Panda D, Radeke MJ, Miller HP, Wilson L, Feinstein SC (1997) Functional interactions between the proline rich and repeat regions of tau enhance microtubule binding and assembly. Mol.Biol.Cell 8: 353–365.

Harlow E, Lane D (1988) Antibodies. A laboratory manual. Cold Spring Harbor Laboratory Publ.

Hirokawa N, Shiomura Y, Okabe S (1988) Tau proteins: the molecular structure and mode of binding on microtubules. J.Cell Biol. 107: 1449–1461.

Hotani H, Horio T (1988) Dynamics of microtubules visualized by darkfield microscopy: treadmilling and dynamic instability. Cell Motil.Cytoskel. 10: 229–236.

Matus A (1988) Microtubule associated proteins. Ann.Rev.Neurosci. 11: 29–44.

Mitchison TJ, Kirschner MW (1984) Dynamic instability of microtubule growth Nature 312: 237–242.

Okabe S, Hirokawa N (1989) Rapid turnover of microtubule-associated protein MAP-2 in the axon revealed by microinjection of biotinylated Map-2 into cultured neurons. Proc. Natl. Acad. Sci. USA 86: 4127–4131.

Olson KR, Olmsted JB (1993) *In vivo* functional analysis of MAP4 Mol. Biol.Cell. 4: 269a.

Wang XM, Peloquin JG, Zhai Y, Bulinski JC, Borisy GG (1996) Removal of MAP4 from microtubules *in vivo* produces no observable phenotype at the cellular level. J.Cell Biol. 132: 345–357.

7 Cytoskeleton Regulation by Rho Proteins

R. Kozma, S. Ahmed and L. Lim

Contents

1   Introduction

The Ras gene encodes a small molecular weight GTPase (p21) which plays a key role in cell growth and differentiation. Mutations in the Ras gene have been found in approximately 30% of all human tumours. Like other G-proteins Ras proteins have intrinsic GTPase and GDP/GTP exchange activity, and cycle between GTP-bound "on" and GDP-bound "off" states. Some Ras oncogenes (e.g. those containing point mutations V12 and L61) encode proteins that are GTPase negative and thereby lock Ras into the GTP-bound "on" state (Feig et al., 1988). In contrast, the mutant Ras N17 protein can suppress oncogenesis and has been termed a dominant negative protein. The N17 mutation changes the nucleotide binding characteristics of the Ras protein, promoting the formation of the GDP-bound state. It is thought that the RasN17 protein suppresses oncogenesis by titrating out a protein that stimulates the intrinsic GDP/GTP exchange activity of Ras, thereby preventing activation of endogenous Ras (Feig and Cooper, 1988).

Methods and Tools in Biosciences and Medicine
Microinjection, ed. by J. C. Lacal et al.
© 1999 Birkhäuser Verlag Basel/Switzerland

The Rho family of p21 GTPases, which includes Rac1, Cdc42Hs and RhoA have >50% amino acid sequence identity with each other and approximately 30% identity with Ras and were first described in the early 1990's. Rho family proteins respond to receptor activation by initiating a program of intracellular changes that includes modification of the actin cytoskeleton, induction of gene expression and control of the cell cycle. Rac1 and Cdc42Hs with V12 and N17, and RhoA with corresponding V14 and N19 point mutations, have similar biochemical characteristics to those described above for Ras and have been used to elucidate protein function. For example, using GTPase negative and dominant negative Rho family proteins, it has been shown that lysophosphatidic acid, platelet derived growth factor and bradykinin activate RhoA, Rac1 and Cdc42Hs to induce the formation of stress fibers, lamellipodia and filopodia, respectively (Ridley and Hall, 1992; Ridley et al., 1992; Kozma et al., 1995). Although the biological function of these structures is not clear, it is thought that stress fibers are involved in cell adherence, lamellipodia in cell movements, and filopodia in sensing environmental cues. Nevertheless, results from these studies suggest that, at least in short-term experiments, the Rho family dominant negative mutant proteins each act by competing for specific exchange factors. This is despite the existence of a large number of potential Rho family exchange factors, which are also often complex multi-domain proteins. As may be expected, the dominant positive mutant proteins each appear to be especially potent at inducing specific changes in morphology. However it should be borne in mind that cycling between GTP and GDP-bound forms of the GTPase could be necessary for long-term actions of these proteins, in which case a dominant positive form would block such a cycle. Here we will describe protein microinjection experiments that have helped to define the functions of Rho family proteins in controlling the actin cytoskeleton of Swiss 3T3 fibroblasts and the morphology of N1E-115 neuroblastoma cells.

2   Materials

Chemicals and items

1,4-Diazabicyclo [2.2.2.] octane	Aldrich	47701
Antibiotic-antimycotic	Gibco	15240–062
Dulbecco's modified Eagle medium	Gibco	41966–029
Femtotips	Eppendorf	5242 952.008
Fetal calf serum	Gibco	16050–080
L-Glutamine	Gibco	25030–024
Microloaders	Eppendorf	5242 956.003
Mowiol	Aldrich	47831
Rhodamine-conjugated phalloidin	Sigma	P1951

Equipment

- Microinjector (Eppendorf, 5242)
- Micromanipulator (Eppendorf, 5170)
- Microscope (Zeiss Axiovert 135)

Solutions

- **Thrombin Cleavage Buffer (Solution 1)**
  50 mM Tris pH 7.6
  50 mM NaCl
  5 mM $MgCl_2$
  5 mM $CaCl_2$
- **GDP-Binding Buffer (Solution 2)**
  50 mM Tris pH7.5
  50 mM NaCl
  5 mM $MgCl_2$
  5 mM DTT
- **Microinjection Buffer (Solution 3)**
  50 mM Tris-HCl pH 7.5
  50 mM NaCl
  5 mM $MgCl_2$
  0.1 mM dithiothreitol
- **Paraformaldehyde solution (Solution 4)**
  Is made by addition of 3% paraformaldehyde to PBS, mixing well, and
  heating to 65°C.
  1 ml 1N NaOH is added per 100 ml of solution to dissolve the fixative.
- **Mowiol (Solution 5)**
  Add 2.4 g of Mowiol 4–88 to 6 g of glycerol
  Add 16 ml of water and mix for a few hours
  Add 12 ml of 0.2 M Tris, pH 8.5
  Heat the mixture to 50°C for 15 min with continual mixing until dissolved
  Add 1,4-diazobicyclo [2.2.2] octane (DABCO) to 2.5%
  Clear by centrifugation at 5000 g for 15 min
  Store aliquots at -20°C and warm at 37°C before use

3   Methods

3.1   Preparation of recombinant rho family GTPases in *E. coli*

Rho family proteins express well as Glutathione-S-transferase (GST)-fusion
proteins in *E.coli* (using the pGEX expression vector system, Pharmacia). From
a liter of *E.coli* culture 1–2 mg of fusion protein can be purified using glu-

tathione-Sepharose chromatography. Rac1 is subcloned into pGEX-2T deriva-
tive p265 (Ahmed et al., 1993, 1994) and Cdc42Hs and RhoA into p265polyG,
derivatives of p265 with a polyglycine spacer between the thrombin cleavage
site and cDNA cloning site (Best et al., 1996). The polyglycine spacer greatly
improves thrombin cleavage of Cdc42Hs and RhoA from GST. GTPase negative
and dominant negative mutants are made using the Clontech Transformer
Site-Directed Mutagenesis kit with the cDNAs in p265 or by first subcloning
them into pBluescript. Mutants are analyzed by DNA sequencing before proteins are expressed in *E. coli*.

3.2   Protein purification

Protocol 1   Protein purification

1.   *E.coli* cells are freshly transformed with the appropriate vector and single
     colonies grown overnight in 20 ml (Luria-broth with ampicillin, 50 mg/ml)
     culture.

2.   The following day the culture is diluted 20–50 fold into 1 litre of Luria-
     broth prewarmed to 37°C.

3.   When the culture reaches an $O.D._{650}$ of 0.5, isopropylthiogalactoside
     (IPTG) is added to 0.5 mM for 1–2 h to allow induction of the fusion pro-
     tein.

4.   *E.coli* cells are harvested by centrifugation at 3000 g, washed once and
     then resuspended in 10 ml of Thrombin Cleavage Buffer **(Solution 1)**.

     Note:
     At this stage the cell extract can be frozen at -20°C and stored.

5.   When required cells are thawed, sonicated (4×30 s, setting 4, MSE sonica-
     tor) and cell debris removed by centrifugation at 14000 g for 10–20 min.

6.   Supernatants are mixed with 1 ml of glutathione-Sepharose resin (Sigma)
     for 10 min in a 50 ml Falcon tube.

7.   The resin is then pelleted by centrifugation at 3000 g for 10 min, the
     supernatant discarded and the resin washed with 10 ml Thrombin Clea-
     vage Buffer three times.

8.   After the final wash the resin is resuspended in 1 ml Thrombin Cleavage
     Buffer, thrombin added (20–40 units, Sigma) and mixing continued for
     10–30 min at room temperature.

9.   The resin is pelleted by centrifugation at 3000 g for 10 min and the super-
     natant (which now contains the p21 protein) mixed with 0.1 ml of benza-
     diamine-Sepharose resin (Sigma) to remove the thrombin.

10. The benzadiamine-Sepharose resin can be removed by centrifugation at 3000 g for 10 min.

11. The resulting supernatant contains the cleaved p21 protein which after concentration is stored at 1 mg/ml or greater at -70°C in aliquots of 20–50 μl.

    The GDP-binding capacity of the Rho family proteins should be determined for each batch of protein:

12. 1 μg p21 protein (1 mg/ml) is incubated with 1 ml of [^{3}H]-GDP (Amersham, 10 Ci/mmol, 1 mCi/ml) in the presence of 10 mM EDTA (to allow exchange) for 10 min at 37°C in 50 ml total reaction volume (with **GDP-binding buffer, Solution 2**).

13. Samples are then filtered on nitrocellulose, washed with 5 ml of ice cold buffer (50 mM Tris pH7.5, 50 mM NaCl) three times and bound counts determined by scintillation counting.

14. The fraction of active protein is estimated by assuming a 1:1 stoichiometry of binding to GDP. The GTPase activity of the p21s can also be measured to characterise the protein batch (Ahmed et al., 1995).

3.3 Cell culture and microinjection

Cell culture
Swiss 3T3 fibroblasts or N1E-115 neuroblastoma cells are grown in Dulbecco's Modified Eagle Media (DMEM) with added L-Glutamine plus 10% fetal calf serum (FCS) at 37°C in 5% CO_2. Both cell lines are grown on either acid-washed glass slides or cell culture grade NUNC plastic dishes. The former should be used if the cells are later to be fixed and stained with fluorescently labelled markers. They can be gridded using a diamond glass marker to allow relocation of microinjected cells.

Microinjection

Protocol 2 Microinjection

1. Sub-confluent cells are starved in DMEM plus 0.2% $NaHCO_3$ (no serum) for 24–48 h before injecting.

   Note:
   In the case of fibroblasts this results in a decline of morphological activity and reduction of stress fibers while in the case of N1E-115 cells it results in neurite outgrowth. The concentration of GTPase protein used varies but is usually in the range of 0.1–2 mg/ml.

2. Proteins are diluted in Microinjection Buffer (**Solution 3**) and cleared by centrifugation (5 min at 14000 g) before use.

3. Protein solution (approx. 1–2 μl) is loaded into glass capillary needles (Femtotips) using Microloader pipets.

4. Microinjection is carried out using an Eppendorf Microinjector (5242) and Micromanipulator (5170) on a Zeiss Axiovert 135 microscope.

5. During microinjection cells are kept at 37°C using a heating stage (TRZ3700), 5% CO_2 using a CO_2 controller (CTI3700), and high humidity maintained by use of a dish of water.

6. Proteins are injected adjacent to the nucleus at approximately 60 hPa for Swiss 3T3 and 100 hPa for N1E-115 cells.

Successful injection can be determined visually at time of injection or by coinjection of a control marker protein such as rat immunoglobulin IgG (2.0 mg/ml) which can later be detected by using fluorescently labelled anti-rat IgG antibodies.

3.4   Determination of actin-based morphology of swiss 3T3 cells

Rhodamine (TRITC)-conjugated phalloidin binds specifically to polymerised filamentous actin (F-actin) and so is very useful to monitor the formation of specific actin polymers. Following microinjection with p21 proteins the primary effects on the cell morphology occur within 1–10 min. From Figure 1 it can be seen that Rho family GTPases induce distinct morphologies based on phalloidin staining. Sub-confluent Swiss 3T3 fibroblasts contain long filamentous stress fibers (Fig. 1a) which can also be induced by RhoA microinjection (Ridley and Hall, 1992). Cdc42Hs microinjection results in the formation of peripheral actin microspikes some of which are filopodia and some retraction fibers (Fig. 1b). Rac1 microinjection results in the formation of membrane ruffles visible as a "curtain" of peripheral actin staining (Fig. 1d). Thirty min after microinjection with Cdc42Hs extensive membrane ruffling is observed (Fig. 1c). Dominant negative p21 (N17 or N19) proteins have been very useful in determining the specificity of these morphological responses. Thus, microinjection with dominant negative Cdc42HsN17 blocks the formation of Cdc42Hs-dependent peripheral microspikes and cellular retraction, and also blocks subsequent membrane ruffling (Fig 1e). In contrast, microinjection with dominant negative Rac1N17 blocks only the formation of membrane ruffling indicating that these structures are Rac1-dependent (Fig 1f). In these assays the dominant negative p21 proteins act in a specific manner. However, their specificity in longer term transfection-based assays is less clear.

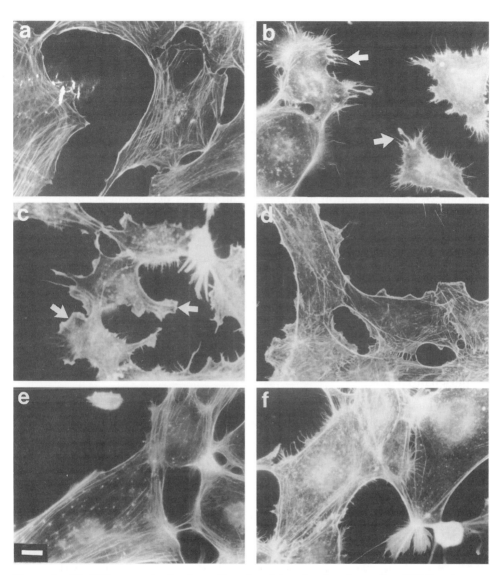

Figure 1   Cdc42Hs promotes the formation of peripheral actin microspikes (PAM) and membrane ruffling in Swiss 3T3 fibroblasts.

Proteins microinjected are as follows: (a) control starved cells, (b) Cdc42Hs (0.5 mg/ml) for 10 min (arrows indicate PAM) and (c) 30 min (arrows indicate areas of ruffling), (d) Rac1 (1 mg/ml) for 10 min, (e) Cdc42Hs (0.5 mg/ml) with Cdc42HsN17 (1 mg/ml) for 10 min, (f) Cdc42Hs (0.5 mg/ml) with Rac1N17 (1 mg/ml) for 30 min. Bar indicates 10 microns.

3.5   Cell fixation and staining with Tritc-Phalloidin

Protocol 3   Cell fixation and staining with Tritc-Phalloidin

1. Rinse cells in phosphate buffered saline (PBS) and fix by adding 3% paraformaldehyde (in PBS, **Solution 4**) and leave at room temperature for 20 min.

2. Remove fixative and rinse in PBS for 10 min.

3. Remove PBS and quench residual fixative by rinsing in 100 mM glycine (in PBS) for 10 min.

4. Remove glycine solution and permeabilize cells by adding 0.2% Triton X-100 (in PBS) for 10 min.

5. Remove Triton solution, rinse briefly in PBS, then block non-specific binding sites by addition of 3% BSA (in PBS) for 10 min.

6. Remove BSA solution then add 1 µg/ml rhodamine-conjugated phalloidin (in PBS with 1% BSA) and leave at 37°C for 40 min.

   Note:
   Care should be taken as phalloidin is a toxic compound.

7. Remove the phalloidin solution and rinse three times with PBS over a total period of 15 min.

8. The slides are then briefly washed in water, allowed to drain for 1 min, and coverslips mounted using Mowiol (**Solution 5**).

9. After the Mowiol has dried, view cells under a fluorescent microscope (Zeiss Axioplan) and photograph with Ektachrome 400 film (Kodak).

4   Procedures and Applications

4.1   Phase-contrast time-lapse analysis of morphological changes

Living cells can be monitored by either phase-contrast or DIC microscopy (we normally use the former) and the time-dependent formation of morphological structures observed. The cells are observed for approximately 10 min to determine the "background" level of morphological activity before they are microinjected with proteins. Using this technique the formation of peripheral Cdc42Hs-dependent filopodia and retraction fibers and Rac1-dependent membrane ruffles, lamellipodia, or intracellular pinocytotic vesicles can be studied in Swiss 3T3 cells (Kozma et al., 1995). However it is not possible to view the formation of the intracellular RhoA-dependent stress fibers in this manner, although RhoA-dependent focal adhesion points where the cell attaches to the substrate can sometimes be seen. In N1E-115 neuroblastoma cells microinjection with Cdc42Hs protein also stimulates the formation of filopodia followed by lamellipodia along the growth cone and neurite (Fig. 2e-h), while Rac1 stimulates formation of lamellipodia only (Fig. 2i-l). Microinjection with RhoA protein induces growth cone and neurite collapse (Fig. 2m-p; Kozma et al., 1997). It cannot be discounted that some form of stress fibers is involved in this RhoA-mediated process, even though normal stress fibers are not easily discernable with phalloidin staining in these cells.

From these short-term injection experiments, it is clear that Cdc42Hs or Rac1 activity results in an increase in growth cone complexity, an activity likely to be important for neurite outgrowth and, *in vivo*, axonal pathfinding. In longer-term injection experiments, neurite outgrowth in these cells can be inhibited by microinjection with dominant negative Cdc42HsN17 or Rac1N17 mutants (the former shown in Figure 3), suggesting that Cdc42Hs and Rac1 are essential for this important developmental process. However, it should be borne in mind that in these neurite outgrowth assays carried out over 24 h the degree of specificity of the response to dominant negative proteins is not yet clear.

Interestingly, there appears to be an antagonistic relationship at least between Cdc42Hs and RhoA in both cell types. Thus, in fibroblasts microinjection with Cdc42Hs results in loss of Rho-dependent stress fibers (Kozma et al., 1995), while in neuroblastoma cells increased growth cone complexity induced by Cdc42Hs can be competed by coinjection with RhoA (Kozma et al.,1997). Withdrawl of serum from these neuroblastoma cells results in neurite outgrowth (Fig. 3A), presumably by removal of factors that activate RhoA since inhibition of endogenous RhoA activity by microinjection with the *C. botulinum* C3 transferase at 0.1 mg/ml also leads to neurite outgrowth in serum-fed cells (Fig. 3C). This activity of C3 transferase can be blocked by coinjection with dominant negative Cdc42HsN17 (Fig. 3D) (as is the effect of serum withdrawl (Fig 3B)) indicating that while RhoA activity appears to prevent neurite outgrowth, Cdc42Hs activity is required for this outgrowth.

For time-lapse studies the cells are maintained on a heated platform (37°C) in a CO_2 box (5%) and viewed under a phase-contrast microscope (Zeiss Axiovert 135) before and after microinjection. Cells can be photographed at time intervals using PAN F 50 film (Kodak). Data can also be collected by use of a video camera (Pulnix TM-6CN) and stored either by a video recorder (Sony u-matic VO-5800PS) or on a computer using image analysis Bio-Rad Comos software. Quantification can be carried out from any of these sources by counting or measuring the extent of particular morphologies. We routinely express morphological activities as a function of time by following cells for a period of 30 min after treatment or microinjection.

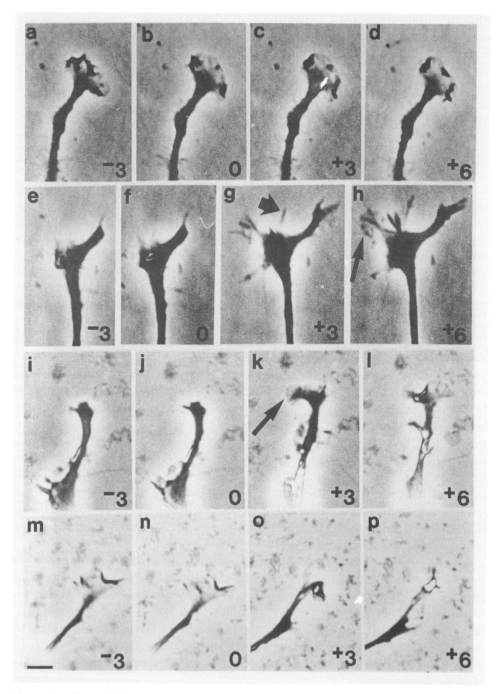

Figure 2 Effects of Cdc42Hs, Rac1, and RhoA on growth cone morphology of neuroblastoma
NIE-115 cells.

Growth cones and neurites of serum starved N1E-115 cells are monitored by phase contrast
microscopy before (first two panels of each horizontal series) and after (last two panels of
each horizontal series) microinjection of proteins. Times in min shown are before (-) and
after (+) microinjection. (c,d) injection with control buffer, (g,h) injection with 1 mg/ml
Cdc42HsV12. Short arrow in panel g indicates a newly formed filopodia; long arrow in panel
h indicates lamellipodia veil formation. (k,l) injection with 1 mg/ml Rac1V12. Arrow in panel
k indicates lamellipodial formation. (o,p) injection with 1 mg/ml RhoA.

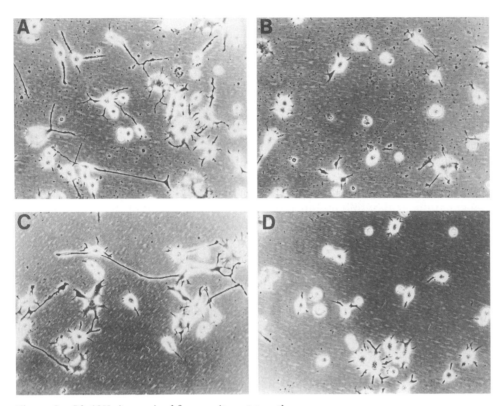

Figure 3   Cdc42Hs is required for neurite outgrowth.

Neuroblastoma cells were serum starved (A) or microinjected with dominant negative Cdc42HsN17 at 2 mg/ml prior to starvation (B). Neurite outgrowth promoted by serum starvation is inhibited by the mutant Cdc42HsN17. In another experiment, cells were grown in serum-containing medium and microinjected with C3 transferase at 0.1 mg/ml (C) or coinjected with C3 transferase plus Cdc42HsN17 at 1 mg/ml (D). Note neurite outgrowth promoted by C3 transferase is again inhibited by the mutant Cdc42HsN17. Cells were microinjected on grids to allow relocation and examined 24 h later by phase contrast microscopy.

4.2   Involvement of rho-family GTPases in factor-driven signalling pathways

The morphological effects in cells following microinjection of Cdc42Hs, Rac1 or RhoA mimic specific effects observed upon stimulation by several physiologically important factors. Pre-injection of cells with the appropriate dominant negative mutant GTPase protein can often specifically block such effects, placing the GTPase within that particular signalling pathway. For instance, in Swiss 3T3 fibroblasts, platelet derived growth factor (10 ng/ml) or epidermal growth factor (10 ng/ml) induce membrane ruffling and lamellipodia formation within 10 min, which are specifically blocked by dominant negative Rac1N17 (Ridley et al.,1992). Dominant negative Cdc42HsN17 specifically blocks filopodial formation and cellular retraction stimulated by 100 ng/ml bradykinin for 10 min (Kozma et al.,1995) and inhibition of RhoA activity blocks stress fiber formation induced by 20 ng/ml lysophosphatidic acid for 10 min (Ridley and Hall 1992). In N1E-115 neuroblastoma cells Cdc42Hs and Rac1-dependent filopodial and lamellipodial activities are stimulated at the growth cone and along the neurite, again within minutes, by the neurotransmitter acetylcholine (Kozma et al. 1997). In this case the agent has to be administered through a micropipet, conditions which create a concentration gradient. In the same way that antagonism between RhoA and Cdc42Hs can be seen at the level of the GTPases themselves, it is also observed at the level of the factors. Thus, in fibroblasts bradykinin-induced stress fiber loss is inhibited by addition of lysophosphatidic acid; while in neuroblastoma cells, neurite retraction induced by lysophosphatidic acid can be competed locally by the administration of acetylcholine through a micropipet (Kozma et al.,1997).

In conclusion, the microinjection of Rho family GTPase proteins into mammalian cells has proven to be a very powerful approach to determine their function, placing them in pivotal roles within signal transduction pathways controlling cell morphology. Future work using these protocols will help to dissect further these important pathways, and to determine more fully the biological function of the actin-based structures being regulated.

Acknowledgements

We thank the Glaxo-Singapore Research Fund for its support.

References

Ahmed S, Lee J, Kozma R et al. (1993) A novel functional target for tumour-promoting phorbol esters and lysophosphatidic acid. J. Biol. Chem. 268: 10709–10712.

Ahmed S, Lee J, Wen L-P et al. (1994) Breakpoint cluster region gene product-related domain of n-chimaerin. Discrimination between Rac-binding and GTPase-activating residues by mutational analysis. J. Biol. Chem. 269: 17642–17648.

Ahmed S, Kozma R, Hall C et al. (1995) GAP activity of n-chimaerin and effect of lipids. Methods in Enzymol. 256: 114–125.

Best A, Ahmed S, Kozma R et al. (1996) The Ras-related GTPase Rac1 binds Tubulin. J. Biol. Chem. 271: 3756–3762.

Feig LA, Cooper GM (1988) Inhibition of NIH 3T3 cell proliferation by a mutant ras protein with preferential affinity for GDP. Mol. Cell. Biol. 8: 3235–3243.

Feig LA, Pan BT, Roberts TM et al. (1988) Relationship among guanine nucleotide exchange, GTP hydrolysis and transforming potential of mutated ras proteins. Mol. Cell. Biol. 8: 2472–2478.

Kozma R, Ahmed S, Best A et al. (1995) The Ras-related protein Cdc42Hs and bradykinin promote formation of peripheral actin microspikes and filopodia in Swiss 3T3 fibroblasts. Mol. Cell. Biol. 15: 1942–1952.

Kozma R, Sarner S, Ahmed S et al. (1997) Rho family GTPases and neuronal growth cone remodelling: relationship between increased complexity induced by Cdc42Hs, Rac1 and acetylcholine and collapse induced by RhoA and lysophosphatidic acid. Mol. Cell. Biol. 17: 1201–1211.

Ridley AJ, Hall A (1992) The small GTP-binding protein rho regulates the assembly of focal adhesions and actin stress fibres in response to growth factors. Cell 70: 389–399.

Ridley AJ, Paterson HF, Johnson C et al. (1992) The small GTP-binding protein rac regulates growth factor-induced membrane ruffling. Cell 70: 401–410.

8 Cytoskeletal Regulation by Rho-p21 Associated Kinases: Analysis of ROK and PAK Function by Plasmid Microinjection

T. Leung, E. Manser and L. Lim

Contents

1   Introduction

Signaling from the Rho-p21 family of small GTP-binding proteins occurs through their interaction with a variety of target proteins, many of which specifically bind the p21s in their active GTP-bound form. Several of these effectors were originally identified in western overlays using [γ^{32}P] labelled Cdc42, Rac1 and RhoA (Manser et al., 1994) while more recently others have been isolated in yeast 2-hybrid screens (Aspenstrom et al., 1995; Watanabe et al., 1996). The ubiquitous p21-activated kinases (PAKs), which were first purified from brain have homologues in yeast worm and flies (recently reviewed by

Methods and Tools in Biosciences and Medicine
Microinjection, ed. by J. C. Lacal et al.
© 1999 Birkhäuser Verlag Basel/Switzerland

Sells and Chernoff, 1997). PAKs bind Cdc42 and Rac1 in their GTP forms but do not interact with RhoA (Manser et al., 1994). Conversely the abundant RhoA-associated kinases (ROKs), which belong to a kinase family with homology to the myotonin kinase, do not bind Cdc42 (Leung et al., 1995). We have recently shown that ROK kinases can evoke many of the cytoskeletal changes associated with RhoA function, including formation of stress fibers and focal adhesion complexes (Leung et al., 1996).

Although in budding yeast the PAK-like kinase Cla4p is implicated in morphological events related to cytokinesis (Cvrckova et al., 1995) there is no evidence that PAK plays such a role in mammalian cells. Rather we have recently shown that PAK is associated with certain focal complexes (phosphotyrosine-rich structures) in both *Drosophila* and mammalian cells (Harden et al., 1996; Manser et al., 1997). Activated versions of PAK cause disassembly of these structures and thus oppose ROK function. *S. cerevisiae* does not appear to contain a ROK-like kinase but rather PKC1 acts as a target kinase for the RhoA homologue Rho1p (Nonaka et al., 1995). Thus although genetics is a powerful tool to dissect the function of kinases in single-cell organisms, the findings do not necessarily extend to multicellular systems. In this chapter we describe the approach of plasmid microinjection into cultured cells to analyze the roles of two Rho-p21 targets, the kinases PAK and ROK. The primary emphasis of these studies relates to morphology of the actin cytoskeleton and the assembly/disassembly of focal contacts (or complexes).

Advantages of plasmid versus protein microinjection
With plasmid microinjection all cDNA constructs can essentially be treated in a similar manner, thus reducing experimental variability. This contrasts with protein micro-injection experiments where each recombinant protein preparation differs in the extent of its purity (and particularly for kinases perhaps the state of activation), and has to be handled accordingly. PAK is a salient example of the problems that can be encountered with the use of recombinant proteins, since when expressed in bacteria the kinase becomes partially activated (probably through auto-phosphorylation). Moreover the kinase activity of PAK is also detrimental to cell growth (Manser et al., 1995, 1997), and expression plasmids harbouring the αPAK cDNA are lethal in certain strains even in the absence of the inducer (ie IPTG).

A common problem in recombinant protein production is the significant amount of proteolysis that occurs in *E. coli*. Thus following one-step purification of Glutathione-*S*-transferase (GST)- or maltose binding protein (MBP)-fusion proteins (most commonly used systems) further purification by chromatography is required. Generally the extent of degradation increases with increased size of the expressed protein, even in protease-deficient strains such as *BL21*. Consequently, although full length PAK (65 kDa) can be recovered from bacteria, full length ROK (160 kDa) cannot. The primary advantage of protein microinjection experiments is the rapidity of cellular effects elicited, occurring within minutes even in the case of the p21 GTPases where C-term-

Mammalian expression is driven by:

CMV enhancer +promoter / β-globin intron (non-coding) which lie 5′ to the epitope tag sequence/MCS/ SV40 polyadenylation signal sequence.

```
    /EcoR1/              <----------- HA Tag --------------> / BamH1/
T7-->GAATTC ACC ATG TAC CCA TAC GAC GTG CCA GAC TAC GCA GGA TCC
               M   Y   P   Y   D   V   P   D   Y   A   G   S

    /EcoR1/              <-------- FLAG Tag -----------> / BamH1/
T7-->GAATTC ACC ATG GAC TAC AAG GAC GAC GAT GAT AAG GGC GGA TCC
               M   D   Y   K   D   D   D   D   K   G   G   S


/ Hind3 / Xho1 /   Not1  /  Sma1 / Pst1 / Sac1  / Kpn1  / Bgl2 /
AAG CTT CTC GAG GCG GCC GCC CGG GCT GCA GGA GCT CGG TAC CAG ATC T
```

Notes: both vector share the same multiple cloning site (MCS)
         the Sac1 site in the multiple cloning site is not unique to the plasmid.
         the BamH1 site is in the same frame as the pGEX series.

Figure 1   Sequences surrounding the multiple cloning site in the pXJ vectors including the epitope sequences recognized by the anti-HA and anti-FLAG Mabs which are described in the text.

These can be used for N-terminal tagging of proteins expressed from the vectors for use in mammalian cells. A T7 polymerase site at the 5′ site allows generation of mRNAs for *in vitro* translation and also sequencing.

inal prenylation is required for biological activity. However the half-life of any particular protein limits its actions in *vivo*. By contrast expression plasmids take ~ 1 h for action, but will maintain high expression levels of most proteins for up to 48 h. An important consideration in DNA microinjection experiments is the type of expression vector to be used, and verification of the protein expressed from it. The next four sections are devoted to these issues; in the last section we present the type of data that can be obtained using plasmid microinjection of various PAK and ROK DNA constructs.

Vectors for mammalian cell microinjection
The pXJ expression plasmid used in all our microinjection studies is derived from that described by Green et al., (1988) which is a high copy number plasmid based on pBluescribe M13+ (Stratagene) and modified to contain a cytomegalovirus (CMV) enhancer and promoter and a SV40 polyadenylation signal downstream of the multiple cloning site (MCS) as described by Xiao et al.

(1991). We have modified this plasmid to include a consensus Kozak translation initiation codon and sequences encoding the haemagglutonin (HA) or FLAG peptide epitopes which lie 5' to the MCS (Fig. 1). These N-terminally positioned "tags" were engineered such that the BamH1 cloning site is in the same frame as that in the pGEX plasmids (Pharmacia), allowing easy inter-conversion of cDNA inserts between the two vectors. Presence of the tags, monoclonal antibodies (Mabs) of which are commercially available, enables detection of the protein without having to produce specific antibodies and allows for direct comparison of the level of expression of different proteins. The pXJ plasmid DNA is purified from XL-1 or JM109 strains of bacteria using Qiagen columns according to the manufacturer's protocol. The plasmid is stored at 0.5 mg/ml in TE buffer (**Solution 1**) at –20°C. Long-term storage of the more dilute plasmid solutions (for microinjection) is not recommended.

2   Materials

Chemicals

Anti-FLAG Mab M2	Kodak/IBI	IB13010
Anti-HA Mab 12CA5	Boehringer Mannheim	1583 816
Anti-vinculin	Sigma	V9131
DNA purification kit	Qiagen	10045
DOSPR transfection reagent	Boehringer Mannheim	14859100
Femtotips	Eppendorf	5242 952
[^{35}S]-methionine	NEN	NEG009A
Rhodamine-phalloidin	Sigma	P1951
PVDF membrane	NEN	NEF 1002
TNT *in vitro* translation kit	Promega	L4610
Lipofectamine	GIBCO-BRL	

Equipment

- Microinjector
- Microscope
- Microloader pipette

Solutions

- **TE buffer (Solution 1)**
  - 10 mM Tris-HCl, pH 8
  - 1 mM EDTA
- **Microinjection buffer (Solution 2)**
  - 50 mM Tris-HCl, pH 7.5
  - 50 mM KCl
  - 5 mM MgCl$_2$

3 Methods

3.1 General methods

Cell culture
A number of cell lines including Swiss 3T3 fibroblasts and HeLa cells show actin-based morphological changes and are suitable for morphological studies by microinjection. We chose HeLa cells for these studies because they are less flattened and more suitable for nuclear injection. HeLa cells are grown in MEM medium containing 10% foetal bovine serum on cover slips gridded with diamond glass marker to allow relocation of microinjected cells. Cells are seeded at low density and used at ~50% confluence after 2–3 days in culture.

DNA preparation
Small particles present in the plasmid DNA samples can potentially block the microinjection pipettes. It is always helpful to remove trace amounts of aggregated material by phenol chloroform extraction (×2), followed by chloroform extraction and ethanol precipitation. After drying, the DNA samples can then be diluted to 0.5 mg/ml with TE buffer (**Solution 1**) for storage at 4°C. For microinjection, samples are clarified by centrifugation (5 min at 14000 g) and further diluted to 50 ng/ml with microinjection buffer (**Solution 2**).

Microinjection
This is carried out using an Eppendorf Microinjector (5242) on a Zeiss Axiovert 35 microscope with a temperature-controlled platform. DNA samples (3 µl) are loaded into a glass capillary needle (Femtotip) using a microloader pipette (Eppendorf). Cultured cells on cover slip are centrally located in the light path and suitable cells in a gridded area chosen. DNA samples are injected into the nucleus at 300 hPa.

3.2 Testing authenticity of the protein product by *in vitro* translation

It is often essential to analyze the size and/or biochemical properties of the resultant protein derived from expression plasmids. The pXJ vectors contains a recognition site for T7 expressed RNA polymerase which allows rapid analysis of the apparent molecular weight of the protein products derived from the vector by coupled *in vitro* transcription/translation as illustrated in Figure 2. The various PAK expression constructs give rise to three closely spaced bands of ~70 kDa arising actually from different translation initiation sites. The largest protein species in each lane are derived from initiation at the ATG flanking the HA tag sequence. Translation is also efficiently driven both from the PAK in-

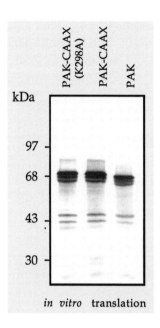

Figure 2 *In vitro* translation products (10 μl equivalent) derived from pXJHA vectors containing the indicated inserts were analyzed on a SDS/polyacrylamide (9%) gel.

The film was exposed overnight. The PAK-CAAX construct encodes PAK fused at the C-terminal side with the last 15 amino acids derived from Cdc42 which includes the CAAX prenylation site. The mutation K298A abolishes kinase activity due to loss of this conserved catalytic lysine (thus behaving as a negative control for microinjection experiments).

itiation codon itself (present in this construct) and at an internal methionine, corresponding to residue 23 of αPAK. Expression *in vivo* however tends to be restricted to the first ATG.

We use the TNT *in vitro* translation system from Promega as follows.

Protocol 1   TNT *in vitro* translation

1. Incubate plasmids (0.5 μg/50 μl reaction) with the modified rabbit reticulolysate mix plus [^{35}S]methionine as instructed for 1 h at 32°C. Include 1 control without added DNA.

2. Dilute each mix with 50 μl of water and 100 μl of 2× SDS sample buffer.

3. Run 40 μl of each reaction mix (without prior heating) on an SDS polyacrylamide gel of appropriate percentage (one lane for protein molecular weight markers).

4. Fix, stain and dry the gel as usual for autoradiography. The signal obtained from the [^{35}S]methionine-labeled protein product should give a clear band after overnight exposure to film; autoradiography enhancers are not required.

If the protein is known to interact with a specific partner, this activity can be tested simply by diluting the *in vitro* translated protein mix in a suitable binding buffer and passing through a 50 μl column of the recombinant protein (partner). Bound protein can be released by heating in SDS sample buffer, and the [^{35}S]methionine-labelled protein detected by autoradiography. This also provides a rapid means of testing mutants which might fail to bind. The authenticity of each cDNA insert (at the 5') can be easily checked by DNA sequencing using a T7 primer.

3.3   Analyzing kinase expression in mammalian cells

To test the *in vivo* stability or activity of the protein product it is necessary to carry out transient transfection of the plasmid in mammalian cells. COS-7 cells grown in 100 mm dishes (Nunc) are routinely used for this purpose.

Protocol 2   Transient transfection of plasmid in mammalian cells

1. Mix 5 mg of plasmid with 40 μl Lipofectamine or DOSPR in 1 ml of serum-free medium in a 15 ml tube.

2. Leave 30 min at room temperature to form the DNA/lipid complex, then add 6 ml of serum-free medium and 1 ml of medium containing 10% FCS.

3. Remove media from the cells and replace with the transfection mix.

4. Leave plates overnight in the CO_2 incubator.

5. The next day remove the media, wash each plate with 10 ml PBS (at room temperature) and scrape each plate of cells on ice into 0.4 ml cell lysis buffer.

6. Transfer lysates to Eppendorf tubes and spin at 18000 g for 10 min at 4°C.

7. Remove the supernatant and measure the protein concentration. Run 80 μg per lane of total protein on a suitable percentage SDS-polyacrylamide gel.

8. Transfer proteins to PVDF membrane under standard conditions for Western analysis, using 1 μg/ml Mab anti-HA (12CA5) or anti-FLAG (M2).

The anti-HA Mab cross-reacts with some endogenous proteins (which will appear in every extract), the strongest of which has an apparent molecular weight of 70 kDa. The signal from the transiently expressed HA-tagged proteins should be at least equal to this band, and usually much stronger. It should not be necessary to expose the Western blot (using luminol) for more than 1 min. The anti-FLAG antibody gives very little backgound signal.

3.4   The *in vivo* activity of protein kinases

Kinases are often subject to complex regulation in the cell and it is therefore not possible to predict whether over-expression will produce active or inactive enzyme. One test of this is to express the kinase by transient transfection and measure the activity of the immuno-precipitated material. In the case of ROK and PAK one can also co-transfect the kinase with its (activating) p21 partner. Using the plasmid construct (for later use in micro-injection experiments) for transient transfection also allows one to test the fidelity of the protein based on the size and activity of the epitope-tagged product. Thus immuno-precipitated PAK has been shown to be essentially inactive (Manser et al., 1997). In the presence of $Cdc42^{G12V}$ *in vitro* PAK undergoes significant activation (Manser et al., 1994) while ROK, which is already active, is only modestly activated by $RhoA^{G14V}$ (Ishizaki et al., 1996). Deletion of non-catalytic regions of kinases can lead to their activation, but may remove important localization signals. A different approach is to generate amino acid substitutions that are activating. In the case of PAK we have shown that substitution of an autophosphorylated threonine in the regulatory loop of the kinase (threonine 422) with an acid residue yields constitutively active kinase (Manser et al., 1997).

Analysis of the activity of various kinase constructs can be carried out using standard immuno-precipitation protocols, using Mabs to the epitope-tagged proteins and analyzing kinase activity using a suitable substrate (with ROK and PAK the common acceptor myelin basic protein is used).

4   Troubleshooting

Apparent lack of expression from plasmid *in vivo*: (1) antibody problems, (2) poor transfection, or (3) insolubility of the protein product.

1. Lack of anti-microbial agent in the stored antibody (use 15 mM sodium azide). Antibodies should not be repeatedly frozen and thawed. Freeze aliquots and working material at 4°C.
2. Ensure that the COS-7 cells are no more than 50–70% confluent at the time of transfection. The plasmid DNA might be impure; make a fresh preparation and always verify presence of the cDNA insert by restriction enzyme digestion. If the plasmid DNA preparation expresses well in transient transfection it is likely to be equally suitable for microinjection.
3. Check the pellet fraction from the 18000 g spin by resuspending in 100 μl of 1× SDS sample buffer, sonicating the sample and using 40 μl for Western analysis.

5   Procedures and Applications

5.1   Detection of expression and cytoskeletal effects

As the constructs are tagged with either HA or FLAG sequences, the expression of the tagged protein is detectable with an anti-tag antibody, often within 1–2 h after injection. The expression level of individual constructs varies, with larger proteins such as ROKα appearing to take a longer time to become detectable (at 2 to 4 h after injection). Filamentous actin is stained with rhodamine phalloidin and anti-vinculin antibody is used to detect focal adhesion complexes.

5.2   Analyzing PAK and ROK function in HeLa cells

To establish the role of a kinase acting downstream of a p21 (or other modulator) it is necessary to obtain a construct which expresses constitutively active kinase. ROK is isolated in an active form, but can be further activated by deletion of the C-terminus containing pleckstrin-homology (PH) and cysteine-rich domains (Leung et al., 1996). A variety of αPAK constructs has been generated which are hyperactive including a construct in which Cdc42 is fused to the C-terminal of PAK, designated αPAK-CC (Manser et al., 1997). All constitutively active versions of PAK exhibit the same phenotype when introduced into HeLa or Swiss 3T3 cells. Figure 3 shows the typical response of cells injected with an expression plasmid encoding αPAK-CC protein, viewed by phase-contrast time-lapse microscopy. Note that 60 min after injection the response of the cells can be clearly seen with the retraction of the cell periphery to leave small processes, while the cell body becomes phase-bright. Injected control cells,

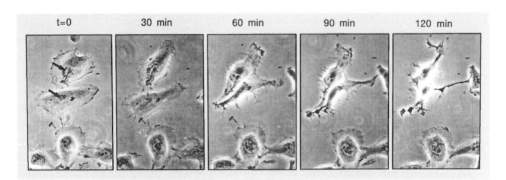

Figure 3   HeLa cells were injected with 50 ng/ml of pXJHA-αPAK-CC plasmid into the nucleus (arrows).

At the indicated times the cells were photographed. Note the retraction of the cell periphery leaving behind small fibers and consequential increase in material around the nucleus which becomes phase bright.

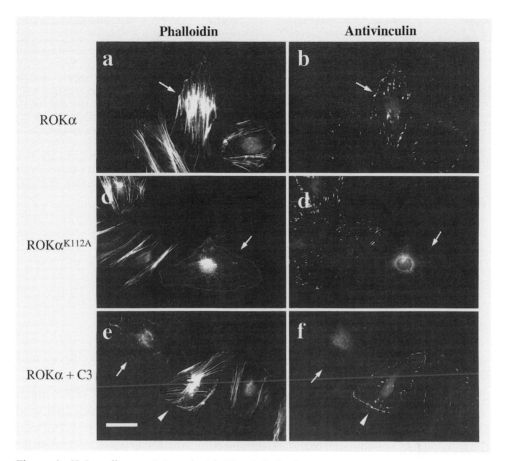

	Phalloidin	Antivinculin
ROKα	a	b
ROKα^K112A	c	d
ROKα + C3	e	f

Figure 4   HeLa cells were injected with 50 ng/ml of either pXJHA-ROKα (a and b) or pXJHA-ROKα^{K112A} plasmid (c and d) into the nucleus (arrows).

In e and f, cells were first injected with C3 transferase (0.2 mg/ml) into the cytoplasm (arrows) followed by injection of pXJ40HA-ROKα into the nucleus 30 min later (arrowheads). Cells were stained with rhodamine-phalloidin (a, c and e) and mouse anti-vinculin/FITC antimouse antibody (b, d and f) 2 h after injection. Images of the stained cells were analyzed with a MRC 600 confocal imager attached to a Zeiss Axioplan fluorescent microscope. Bar=10 μm.

containing vector plasmid alone or plasmid encoding kinase inactive PAK-CC(K298A), do not exhibit such changes over this time period. The inability of the construct encoding inactive PAK mutant to promote similar effects is strong evidence that these morphological changes occur as a consequence of PAK phosphorylation of target proteins. As discussed in more detail (Manser et al., 1997) the cell retraction probably results directly from the ability of PAK to disassemble focal adhesion complexes and actin stress fibers.

In contrast to the effects of PAK, expression of the ROKα kinase leads to an increase in the stress fiber content and in focal adhesion complexes. Figure 4 shows the effects of wild-type and kinase dead (K112A) ROKα on filamentous

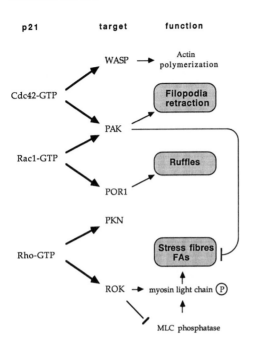

p21   target   function

Figure 5   Involvement of Rho-p21 targets in cell morphology.

The role of some of the identified targets of the Rho-p21s are shown. Experiments with constitutively active PAK suggest that it plays a role both in cell retraction (a Cdc42 phenotype) and in opposing the actions of RhoA. The ROKs or Rho-kinases which induce formation of stress fibers and focal adhesions (FAs) are known to promote phosphorylation of myosin light chain by the two pathways shown. WASP, Wiskott-Aldrich syndrome protein; POR 1, partner of Rac1; PKN, protein kinase N.

actin and focal adhesion complexes. Enhanced formation of stress fibers and focal adhesion complexes is observed when plasmids encoding wild-type ROKα (a and b) are injected (arrows). By contrast a "dominant negative" effect involving loss of stress fibers and focal adhesions is observed in cells injected with plasmid encoding the inactive kinase mutant (Fig. 4c and d). The *C. botulinum* C3 transferase specifically ADP-ribosylates Rho and abolishes Rho action including promotion of stress fibers and focal adhesion complexes (arrowed in Fig. 4e and f). However, cells co-injected with C3 transferase and ROKα plasmid maintain these structures (Fig 4e and f; arrowheads), suggesting that ROKα acts downstream of Rho in cytoskeletal organization. These observations with ROK and PAK, in relation to other known p21 "targets" are summarized in Figure 5. The effects of ROK may be mediated both by direct phosphorylation of the myosin light chain and through inhibition of the myosin light chain (MLC) phosphatase, thereby increasing the contractility of actin/myosin complexes and leading to stress fiber formation. This may be sufficient to drive the formation of focal adhesions. As yet the signalling components that lie "downstream" of Cdc42 and Rac1 are less clearly defined. Experiments using the type of methodology described in this chapter will no doubt soon yield new information on these.

Acknowledgements

We thank the Glaxo-Singapore Research Fund for its support.

References

Aspenstrom P, Lindberg U, Hall A (1995) Two GTPases, Cdc42 and Rac, bind directly to a protein implicated in the immunodeficiency disorder Wiskott-Aldrich syndrome. Current Biol. 6: 70–75.

Cvrckova F, De Virgilio C, Manser E et al. (1995) Ste20-like protein kinases are required for normal localization of cell growth and for cytokinesis in budding yeast. Genes & Dev. 9; 1817–1830.

Green S, Issemann I, Sheer E (1988) A versatile in vivo and in vitro vector for protein engineering. Nucleic Acid Res. 16: 369.

Harden N, Lee J, Loh H-Y et al. (1996) A Drosophila homolog of the Rac- and Cdc42-activated serine/ threonine kinase PAK is a potential focal adhesion and focal complex protein that co-localizes with dynamic actin structures. Mol. Cell. Biol. 16: 1896–1908.

Ishizaki T, Naito M, Fujisawa K et al. (1997) p160^{ROCK}, a Rho-associated coiled-coil forming protein kinase, works downstream of Rho and induces focal adhesions. FEBS Letts. 404: 118–124.

Leung T, Manser E, Tan L et al. (1995) A novel serine/threonine kinase binding the Ras-related RhoA GTPase which translocates the kinase to peripheral membranes. J. Biol. Chem. 270: 29051–29054.

Leung T, Chen X-Q, Manser E et al. (1996) The p160 RhoA-binding kinase ROKα is a member of a kinase family and is involved in the reorganization of the cytoskeleton. Mol. Cell. Biol. 16: 5313–5327.

Manser E, Leung T, Salihuddin H et al. (1994) A brain serine/threonine protein kinase activated by Cdc42 and Rac1. Nature 367: 40–46.

Manser E, Chong C, Zhao Z-S et al. (1995) Molecular cloning of a new member of the p21-Cdc42/Rac-activated kinase (PAK) family. J. Biol. Chem. 270: 25070–25078.

Manser E, Huang H-Y, Loo T-H et al. (1997) Expression of constitutively active α-PAK reveals effects of the kinase on actin and focal complexes. Mol. Cell. Biol. 17: 1129–1143.

Nonaka H, Tanaka K, Hirano H et al. (1995) A downstream target of RHO1 small GTP-binding protein is PKC1, a homolog of protein kinase C, which leads to activation of the MAP kinase cascade in Saccharomyces cerevisiae. EMBO J. 14: 5931–5938.

Sells MA, Chernoff J (1997) Emerging from the Pak: the p21-activated protein kinase family. Trends Cell Biol. 7: 162–167.

Watanabe G, Saito Y, Madaule P et al. (1996) Protein kinase N (PKN) and PKN-related protein Rhophilin as targets of the small GTPase Rho. Science 271: 645–648.

Xiao JH, Davidson I, Matthes H et al. (1991) Cloning, expression and transcriptional properties of the human enhancer factor TEF-1. Cell 65: 551–568.

9 Use of Mircoinjection to Study Apoptosis and its Prevention

F. Kohlhuber, H. Hermeking and D. Eick

Contents

1 Introduction

Programmed cell death is an intrinsic death program operating to remove un-wanted cells during embryonal development. It has also been suggested to eliminate cells which have acquired growth factor-independent growth prop-

Methods and Tools in Biosciences and Medicine
Microinjection, ed. by J. C. Lacal et al.
© 1999 Birkhäuser Verlag Basel/Switzerland

erties due to genetic alterations (Evan and Littlewood, 1993). Common morphological features of programmed cell death are blebbing of the plasma membrane, chromatin condensation, fragmentation of the nucleus, and breaking up of the cell into apoptotic bodies (Wyllie, 1980). Apoptotic cell death involves a cell-intrinsic program that can operate in the presence of inhibitors of RNA and protein synthesis. In some cellular systems, it is even induced by these inhibitors, indicating that short-lived proteins or RNAs may control the apoptotic machinery negatively (Ellis et al., 1991).

Cellular and viral oncogenes have been implicated in the control of proliferation and apoptosis (Eick and Hermeking, 1996). Constitutive expression of the proto-oncogene c-*myc*, for example, prevents cell cycle exit of mouse fibroblasts after withdrawal of growth factors. These cells continue cycling and concomitantly undergo apoptosis (Evan et al., 1992; Hermeking and Eick, 1994). Similar to c-*myc*, a number of viral oncogenes can activate the cell cycle. Cell lines expressing certain viral or cellular oncogenes are, therefore, probably subject to a genetic selection against apoptosis, making the study of apoptosis difficult in stable cell lines. Conditional expression of viral and cellular oncogenes might be a solution to this problem. This allows the selection of cell lines in the absence of oncogene activity. Alternatively, as demonstrated in the following section, apoptosis can be studied in microinjection experiments. We show that serum-starved mouse fibroblasts undergo apoptosis after microinjection of expression plasmids coding for cellular and viral oncogenes and that co-expression of survival factors or the inactivation of the tumour suppressor gene product p53 can inhibit apoptosis. Microinjection is, therefore, a suitable method to study the apoptotic and anti-apoptotic potential of gene products.

2 Materials

Chemicals and cell culture

4`,6-diamidino-2-phenylindole (DAPI)	Sigma
5-bromo-2'-deoxyuridine (BrdU)	Amersham
Acetone	Merck
AMCA-coupled donkey F(ab)2` anti-rabbit IgG	Dianova, Hamburg, Germany
Anti-Bcl-2 antibody (mouse monoclonal)	kindly provided by D.Y. Mason, Headington, U.K.
Anti-BrdU-Fluorescein antibody (mouse monoclonal)	Boehringer Mannheim
Anti-c-myc antibody (9E10; mouse monoclonal)	Boehringer Mannheim
Anti-hemagglutinin antibody (12CA5, mouse monoclonal)	Boehringer Mannheim, Germany
Anti-SV 40 large T-antigen (K142, rabbit polyclonal)	kindly provided by E. Fanning, Nashville, TN, USA
Bovine Serum Albumin (BSA), fraction V, protease-free	Sigma
Dulbecco`s Modified Eagle`s Medium (DMEM)	Gibco/BRL Life Technologies

Fetal Calf Serum (FCS)	Serva
Glycerol	Merck
HCl	Merck
Methanol	Merck
Paraformaldehyde	Sigma
Streptavidin-Cy3	Dianova, Hamburg, Germany
TRITC-coupled goat F(ab)2` anti-rabbit IgG	Dianova, Hamburg, Germany
TRITC-coupled goat IgG anti-mouse IgG	Dianova, Hamburg, Germany
Triton X-100	Sigma

Antibodies: supernatant of hybridoma cell lines was used undiluted or diluted up to five-fold; commercial antibodies were diluted in the range from 1:50 to 1:2000.

3T3-L1 mouse fibroblasts were cultivated in Dulbecco's modified Eagle's medium (DMEM) supplemented with 10% fetal calf serum (FCS) in a humidified atmosphere containing 5% carbon dioxide at 37°C.

Equipment

- Capillary puller, Kopf Instruments, Tujunga, CA, USA
- CCD-camera, Hitachi
- Cellocate (175 µm), Eppendorf-Netheler-Hinz, Hamburg, Germany
- DNA-preparation columns, Qiagen
- Glass capillaries (1.2 mm, with inner filament), Clark Electro Medical Instruments, U.K.
- Microinjector 5242, Eppendorf-Netheler-Hinz, Hamburg, Germany
- Microloader, Eppendorf-Netheler-Hinz, Hamburg, Germany
- Micromanipulator 5171, Eppendorf-Netheler-Hinz, Hamburg, Germany
- Microscope Axioscop, Zeiss, Oberkochen, Germany
- Microscope Axiovert 10, Zeiss, Oberkochen, Germany
- Microscope camera MC100, Zeiss, Oberkochen, Germany
- Monitor, Hitachi
- Tissue culture plates (35 mm), Nunc, Wiesbaden, Germany

Plasmids

Simian virus 40 (SV40) large tumor antigen (LTAg) and two mutants, LTAg-K1 and LTAg-(1–259), were expressed under the control of the SV40 early promoter (Hermeking et al., 1994). Wild-type c-myc, the c-myc mutant ctMycBR⁻ and the Max mutant MaxBR⁻ were expressed under the control of the cytomegalovirus (CMV) promoter/enhancer (Hermeking et al., 1994; Kohlhuber et al., 1995). Wild-type c-myc and MaxBR⁻ were tagged at the carboxy-terminus with a viral hemagglutinin (HA) epitope which is recognized by the monoclonal antibody 12CA5. For plasmids see Figure 1.

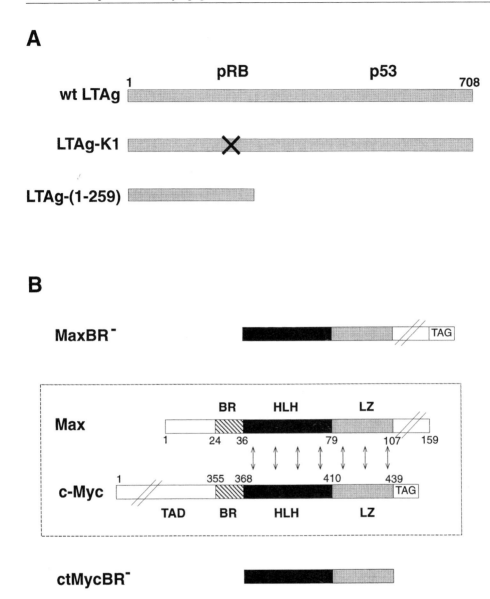

Figure 1 Schematic overview of wild-type LTAg, c-myc, Max and the corresponding mutants.

(A) Wild-type SV40 LTAg (708 amino acids) is shown in the upper half (Dobbelstein et al., 1992). In the mutant LTAg-K1 which is unable to bind pRB, E_{107} has been replaced by K (Kalderon and Smith, 1984). Mutant LTAg-(1–259) is a deletion mutant unable to bind p53 (Dobbelstein et al., 1992). (B) The c-myc protein (439 amino acids) has features of a transcription factor with a transcriptional transactivation domain (TAD) in the amino-terminal region. The carboxy-terminus harbors a basic region (BR) required for DNA-binding and a helix-loop-helix/leucine zipper motif (HLH-LZ) required for dimerization with the Max protein. c-myc/Max heterodimers bind to DNA and modulate the transcription of target genes. ctMycBR⁻ and MaxBR⁻ have a deletion of their amino-termini up to the HLH-domain and can act as dominant negative mutants.

Solutions

- **Phosphate Buffered Saline (PBS)**
    Dissolve in 800 ml of distilled water:
    8.0 g of NaCl
    0.2 g of KCl
    1.44 g of Na_2HPO_4
    0.24 g of KH_2PO_4
    Adjust the pH to 7.2
    Adjust the volume to 1 liter
- **Tris/EDTA (TE), pH 7.5**
    10 mM Tris-HCl, pH 7.5
    1 mM EDTA.
- **4% Paraformaldehyde**
    Add 8.0 g paraformaldehyde to 100 ml of water
    Heat to 60°C in a fume hood
    Add a few drops of 1 N NaOH to help dissolve
    When the solid has completely dissolved, let the solution cool to room temperature
    Add 100 ml 2× PBS
- **Permeabilisation solution**
    0.2% Triton-X100 in PBS
- **Blocking solution**
    3% bovine serum albumin (BSA) or, alternatively, 10% FCS
    in PBS containing 0.01% Na-azide

3   Methods

3.1   Microinjection of an expression plasmid coding for SV40 LTAg induces cellular DNA synthesis in quiescent 3T3-L1 fibroblasts

Microinjection of purified SV40 LTAg triggers DNA synthesis in quiescent cells (Tjian et al., 1978). Similar results were obtained when expression plasmids coding for LTAg were microinjected into serum-starved mouse 3T3-L1 cells (Fig. 2) (Hermeking et al., 1994).

Serum-starvation of 3T3-L1 cells

Protocol 1   Serum-starvation of 3T3-L1 cells

Day 1 Plate 1.5×10^5 cells on glass coverslips with a system of coordinates (Cellocate) in 35-mm tissue culture plates in 3 ml DMEM/10% FCS.

Day 2 The cells should be semi-confluent. Wash cells three times in DMEM only (without serum) for 1 min each. After the last wash, incubate cells in DMEM only for 15 min at 37°C. Remove DMEM and incubate cells in 4.5 ml DMEM/0.5% FCS for 48 h.

Day 4 The cells are now ready for microinjcetion.

Microinjection of an expression plasmid coding for LTAg

Protocol 2   Microinjection of an expression plasmid coding for LTAg

1. Plasmid DNA was prepared using Qiagen columns and DNA was redissolved in TE, pH 7.5, at a concentration of 50 ng/ml.

2. Load DNA into injection needles using Femtotips (Eppendorf).

3. DNA is injected into the nucleus with an injection pressure of 20–50 hPa and an injection time of ~0.3 s.

   Note:
   Approximately 5–30% of the microinjected cells do not survive the procedure of microinjection. Cell death that occurs immediately after microinjcetion is probably due to damaging of the cell membrane. The damaged cells die within 5–10 min after microinjection and do not show characteristic features of apoptosis. They lift off the glass coverslip within 30 min after microinjection.

4. BrdU labelling solution (1 μl/ml) is added to each dish 14 h after microinjection and the incubation is continued for another 6 h at 37°C.

Fixation, staining and immunofluorescence

Protocol 3   Fixation, staining and immunofluorescence

1.   Wash cells three times in PBS.

2.   Remove PBS and fix cells in ice-cold paraformaldehyde for 20 min at 4°C.

3.   Remove paraformaldehyde and wash cells 3 times in PBS.

4.   Permeabilize cell membranes by incubating in permeabilization solution for 20 min at room temperature.

5.   Wash cells three times in PBS and incubate in blocking solution for at least 1 h at 4°C.

6.   Remove blocking solution, add the primary antibody (rabbit anti-LTAg antibody diluted in blocking solution) and incubate for 1 h at room temperature in a humidified chamber.

7.   Wash cells in three changes of PBS over 5 min, add the secondary antibody (TRITC-coupled anti-rabbit antibody diluted in blocking solution) and incubate for 1 h at room temperature.

8.   Wash cells in three changes of PBS over 5 min. For detection of incorporated BrdU, the cellular DNA is now denatured by incubating in 2 N HCl for 15 min at 37°C.

9.   Wash cells in three changes of PBS over 5 min, add FITC-coupled anti-BrdU antibody (diluted in blocking solution) and incubate for 30 min at 37°C.

10. Wash cells in three changes of PBS over 5 min and mount in 70–80% glycerol containing DAPI (0.0005%) to stain nuclei. The absorption and emission curve of DAPI does not overlap the narrow band filters for fluorescein and rhodamine.

In our experiments, almost all cells that expressed LTAg had also incorporated BrdU (>95%) (Fig. 2). The background of cells that incorporated BrdU without expressing LTAg was consistently less than 3%. With the exception of those cells which died due to membrane damage immediately after microinjection, no further cell death was observed after expression of LTAg in serum-starved cells. A mutant of LTAg that is no longer able to bind the retinoblastoma suppressor gene product pRB, LTAg-K1 (Fig. 1) did not induce DNA-synthesis (data not shown).

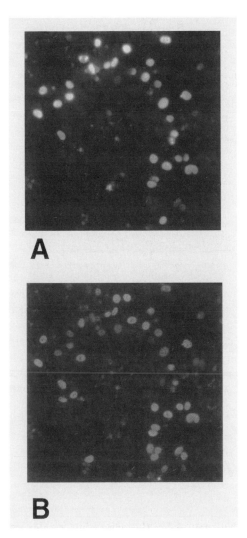

Figure 2 LTAg induces DNA-synthesis in serum-starved 3T3-L1 fibroblasts.

Serum-starved cells were microinjected with an expression plasmid coding for LTAg (50 ng/ml). DNA-synthesis was measured by incorporation of BrdU. 20 h after microinjection, cells were fixed and analyzed by immunocytochemistry: (A) LTAg: red, (B) BrdU: green (magnification ×200).

3.2 Microinjection of expression plasmids coding for c-myc and a c-myc mutant induces apoptosis in quiescent 3T3-L1 fibroblasts

Similar to LTAg, the expression of the proto-oncogene c-myc has been reported to induce DNA synthesis in serum-starved mouse fibroblasts. The expression vector for wild-type c-myc (Fig. 1) was microinjected into serum-starved 3T3-L1 cells. The cells were fixed 20 h after microinjection and stained with anti-c-myc and anti-BrdU antibodies. In these experiments no living cells expressing c-myc could be detected (Hermeking et al., 1994). Almost all of the cells in the

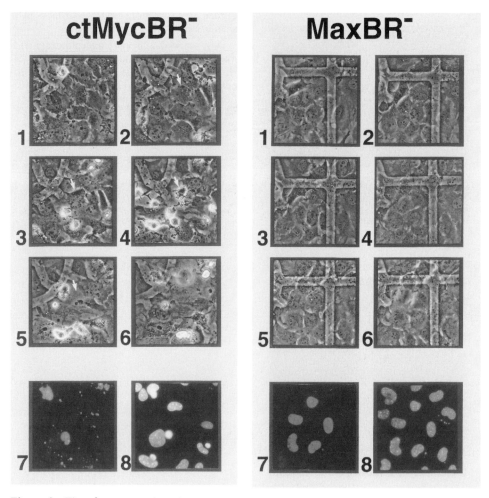

Figure 3   Time-lapse experiment.

Serum-starved 3T3-L1 cells were microinjected with expression plasmids coding for MaxBR⁻ and ctMycBR⁻ (150 ng/ml). The field of injected cells was photographed every 2 h (1–6). 12 h after microinjection, the cells were fixed and stained with anti-c-myc or anti-HA antibodies (red fluorescence, 7) and DAPI (blue, 8) (magnification ×400).

injected field were disintegrated and most of them detached from the glass coverslip. Fixing and staining cells 2 h after microinjection, however, showed a bright fluorescence with the anti-c-myc antibody and the cells looked normal. Similar results were obtained with the deletion mutant ctMycBR⁻ (Fig. 1) (Kohlhuber et al., 1995).

We performed time-lapse experiments to be able to follow the events in cells expressing ctMycBR⁻ (Fig. 3). For this purpose, the injected cells were inspected and photographed in intervals of 2 h after microinjection. Expression of the mutant could be detected 2 h after microinjection in more than 70% of

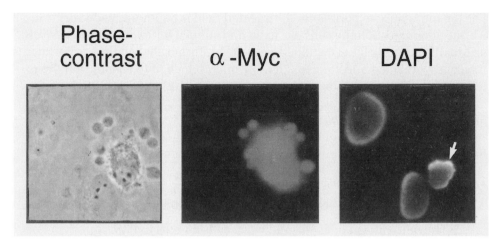

Figure 4   Apoptotic cell expressing ctMycBR⁻.

Serum-starved cells were microinjected with an expression plasmid coding for ctMycBR⁻ (150 ng/ml). 6 h after microinjection, the cells were fixed and stained with anti-c-myc antibody (red fluorescence) and DAPI (blue). Cytoplasmic blebbing can clearly be seen in the phase contrast, whereas chromatin condensation can be seen in the DAPI-stained nucleus (indicated by an arrow) (magnification ×1000).

the injected cells (data not shown). At that time point the cells looked normal and showed a regular blue chromatin staining with DAPI. Four h after microinjection, however, some of the injected cells started changing their morphology and bright spots became apparent in the field of injected cells. The number of affected cells consistently increased between 4 to 12 h. After 12 h, cells staining with the c-myc antibody and showing a normal shape were no longer detectable. Close examination including staining of the nuclei with DAPI indicated that the microinjected cells died of apoptosis: we observed cytoplasmic blebbing, nuclear fragmentation and breaking up of the cells into apoptotic bodies (Fig. 4). In contrast to nuclei of non-apoptotic cells, chromatin staining (DAPI) of apoptotic nuclei was irregular with bright areas which is indicative of chromatin condensation. In a similar time-lapse experiment, apoptosis was not observed in cells microinjected with an expression plasmid coding for an amino-terminal deletion mutant of Max (MaxBR⁻, Figs 1 and 3).

Serum-starvation of 3T3-L1 cells

See **Protocol 1**.

Microinjection of an expression plasmid coding for ctMycBR⁻ and MaxBR⁻

See **Protocol 2**, 1.-3., and optional 4 if one wants to look at DNA-synthesis.

For the time-lapse experiments the injected field is photographed every 2 h; 12 h after microinjection the cells are fixed and stained for expression of ctMycBR⁻ or MaxBR⁻, the nuclei are stained with DAPI. In parallel control experiments the cells were fixed and stained 2 h after microinjection to ensure that the proteins are expressed at a time point were apoptosis has not yet started.

Fixation, staining and immunofluorescence

Protocol 4 Fixation, staining and immunofluorescence

1. Wash cells three times in PBS.

2. Remove PBS and fix cells in ice-cold paraformaldehyde for 20 min at 4°C.

3. Remove paraformaldehyde and wash cells 3 times in PBS.

4. Permeabilize cell membranes by incubating in permeabilization solution for 20 min at room temperature.

5. Wash cells three times in PBS and incubate in blocking solution for at least 1 h at 4°C.

6. Primary antibodies: anti-c-myc antibody (9E10) for the detection of ctMycBR⁻, anti-HA antibody (12CA5) for the HA-tagged Max protein.

7. Wash cells in three changes of PBS over 5 min, add the secondary antibody (TRITC-coupled anti-mouse antibody diluted in blocking solution) and incubate for 1 h at room temperature.

8. Wash cells in three changes of PBS over 5 min. For detection of incorporated BrdU, the cellular DNA is now denatured by incubating in 2 N HCl for 15 min at 37°C.

9. Wash cells in three changes of PBS over 5 min, add FITC-coupled anti-BrdU antibody (diluted in blocking solution) and incubate for 30 min at 37°C.

10. Optional (staining for BrdU): Wash cells in three changes of PBS over 5 min and mount in 70–80% glycerol containing DAPI (0.0005%) to stain nuclei. The absorption and emission curve of DAPI does not overlap the narrow band filters for fluorescein and rhodamine.

3.3   C-myc-induced apoptosis is suppressed by co-expression of LTAg

C-myc-induced apoptosis in serum-straved 3T3-L1 fibroblasts has been shown to require the tumour suppressor protein p53 (Hermeking and Eick, 1994). p53 is involved in a subset of apoptotic responses and its inactivation by muta-

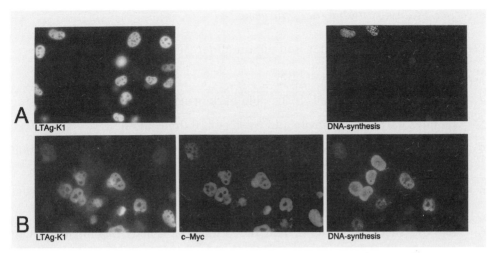

Figure 5   C-myc-induced apoptosis is suppressed by co-expression of LTAg-K1.

Serum-starved 3T3-L1 cells were microinjected with (A) an expression plasmid coding for LTAg (50 ng/ml) or (B) a mixture of expression plasmids coding for LTAg (50 ng/ml) and c-myc (25 ng/ml). DNA-synthesis was measured by incorporation of BrdU. The cells were fixed 20 h after microinjection and stained for LTAg (blue), c-myc (red) and BrdU (green) (magnification ×400).

tion or viral oncoproteins can lead to resistance against apoptosis. Both LTAg and LTAg-K1 are capable of binding and inactivating p53 (Fig. 1). We therefore asked whether inactivation of p53 by expression of LTAg-K1 inhibits c-myc-induced apoptosis (Fig. 5). When c-myc and LTAg-K1 were co-expressed in quiescent 3T3-L1 cells the frequency of cells staining positively for c-myc was increased from <3% to more than 30% after 20 h, indicating that c-myc-induced apoptosis could indeed be suppressed by LTAg-K1. In addition, almost all cells that stained positive for LTAg-K1 and c-myc also stained positive for BrdU, demonstrating that the defect of LTAg-K1 in induction of DNA-synthesis was restored by expression of c-myc.

Serum-starvation of 3T3-L1 cells

See **Protocol 1.**

Microinjection of an expression plasmid coding for c-myc and LTAg-K1

See **Protocol 2**; serum-starved cells were microinjected with expression plasmids encoding LTAg-K1 (50 ng/ml) alone (Fig. 5A), or with a combination of LTAg-K1 (50 ng/ml) and c-myc (25 ng/ml) (Fig. 5B).

Fixation, Staining and Immunofluorescence

Protocol 5   Fixation, staining and immunofluorescence

1.   Wash cells three times in PBS.

2.   Remove PBS and fix cells in ice-cold paraformaldehyde for 20 min at 4°C.

3.   Remove paraformaldehyde and wash cells 3 times in PBS.

4.   Permeabilize cell membranes by incubating in permeabilization solution for 20 min at room temperature.

5.   Wash cells three times in PBS and incubate in blocking solution for at least 1 h at 4°C.

6.   Remove blocking solution, add the primary antibody (rabbit anti-LTAg antibody diluted in blocking solution) and incubate for 1 h room temperature in a humidified chamber.

7.   Wash cells in three changes of PBS over 5 min, add the secondary antibody (AMCA-coupled anti-rabbit antibody diluted in blocking solution) and incubate for 1 h at room temperature.

8.   Wash cells in three changes of PBS over 5 min, add the biotinylated 12CA5 antibody (directed against the HA-tagged c-myc protein) and incubate for 1 h at room temperature.

9.   Wash cells in three changes of PBS over 5 min, add Streptavidin-Cy3 (diluted in blocking solution) and incubate for 1 h at room temperature.

10. Wash cells in three changes of PBS over 5 min. For detection of incorporated BrdU, the cellular DNA is now denatured by incubating in 2 N HCl for 15 min at 37°C.

11. Wash cells in three changes of PBS over 5 min, add FITC-coupled anti-BrdU antibody (diluted in blocking solution) and incubate for 30 min at 37°C.

12. Wash cells in three changes of PBS over 5 min and mount in 70–80% glycerol containing DAPI (0.0005%) to stain nuclei. The absorption and emission curve of DAPI does not overlap the narrow band filters for fluorescein and rhodamine.

3.4   Bcl-2 inhibits apoptosis induced by a truncated form of LTAg

Analysis of various mutants of LTAg revealed that the binding domain of p53 is required to inhibit c-myc-induced apoptosis. A carboxy-terminal deleted mutant of LTAg, LTAg-(1–259) (Fig. 1), is unable to bind p53 and does not inhibit

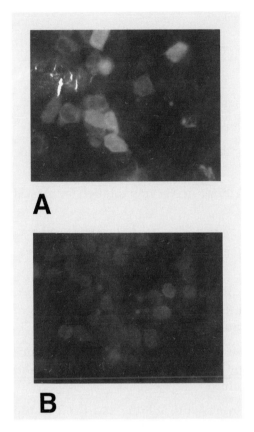

Figure 6 Bcl-2 suppresses LTAg-(1–259)-induced apoptosis.

Serum-starved 3T3-L1 cells were microinjected with a mixture of expression plasmids coding for LTAg-(1–259) (50 ng/ml) and Bcl-2 (25 ng/ml). DNA-synthesis was measured by incorporation of BrdU. The cells were fixed 20 h after microinjection and stained for (A) LTAg (blue, nuclear staining), Bcl-2 (red, cytoplasmic staining) and (B) BrdU (green) (magnification ×400).

c-myc-induced apoptosis. Instead, this mutant induced apoptosis to a similar extent as c-myc after expression in serum-starved fibroblasts (data not shown).

The cellular proto-oncogene *bcl-2* has previously been shown to inhibit c-myc-induced apoptosis (Fanidi et al., 1992). Here we show that expression of Bcl-2 can also inhibit apoptosis induced by LTAg-(1–259) (Fig. 6).

Serum-starvation of 3T3-L1 cells

See **Protocol 1**.

Microinjection of an expression plasmid coding for LTAg-(1–259) and Bcl-2

See **Protocol 2**; serum-starved cells were microinjected with a combination of expression plasmids encoding LTAg-(1–259) (50 ng/ml) and Bcl-2 (25 ng/ml).

Fixation, staining and immunofluorescence

Protocol 6   Fixation, staining and immunofluorescence

1. Wash cells three times in PBS.

2. Remove PBS and fix cells in ice-cold paraformaldehyde for 20 min at 4°C.

3. Remove paraformaldehyde and wash cells 3 times in PBS.

4. Permeabilize cell membranes by incubating in permeabilization solution for 20 min at room temperature.

5. Wash cells three times in PBS and incubate in blocking solution for at least 1 h at 4°C.

6. Remove blocking solution, add the primary antibody (rabbit anti-LTAg antibody diluted in blocking solution) and incubate for 1 h room temperature in a humidified chamber.

7. Wash cells in three changes of PBS over 5 min, add the secondary antibody (AMCA-coupled anti-rabbit antibody diluted in blocking solution) and incubate for 1 h at room temperature.

8. Wash cells in three changes of PBS over 5 min, add the secondary antibody (AMCA-coupled anti-rabbit antibody diluted in blocking solution) and incubate for 1 h at room temperature.

9. Wash cells in three changes of PBS over 5 min, add the anti-Bcl-2 mouse monoclonal antibody (diluted in blocking solution) and incubate for 1 h at room temperature.

10. Wash cells in three changes of PBS over 5 min, add the secondary antibody (TRITC-coupled anti-mouse antibody diluted in blocking solution) and incubate for 1 h at room temperature.

11. Wash cells in three changes of PBS over 5 min. For detection of incorporated BrdU, the cellular DNA is now denatured by incubating in 2 N HCl for 15 min at 37°C.

12. Wash cells in three changes of PBS over 5 min, add FITC-coupled anti-BrdU antibody (diluted in blocking solution) and incubate for 30 min at 37°C.

13. Wash cells in three changes of PBS over 5 min and mount in 70–80% glycerol containing DAPI (0.0005%) to stain nuclei. The absorption and emission curve of DAPI does not overlap the narrow band filters for fluorescein and rhodamine.

4   Troubleshooting

Cells die immediately after microinjection
- The plasmid-solution is contaminated (e.g. salt, phenol, etc.): try other protocols for plasmid preparation.
- The cells are damaged by the injection needles: try different pulling protocols or use commercially available needles (e.g. Femtotips, Eppendorf).
- The injection pressure is too high and/or the injection time is too long: adjust both for your specific cell line.

Immunofluorescence staining does not work
- The antibodies are diluted to high (no or only weak staining) or to low (high background): make a titration of the primary and secondary antibodies.
- Fixing reagent is not compatible with epitope: try other fixing reagents (e.g. acetone/methanol instead of paraformaldehyde).
- The promoter/enhancer does not work in your particular cell line: clone your gene of interest into a different expression plasmid.

Acknowledgements

Supported by the Deutsche Forschungsgemeinschaft (SFB190: Mechanismen der Genaktivierung), the Fonds der Chemischen Industrie and the Deutsche Krebshilfe.

References

Dobbelstein M, Arthur AK, Dehde S, van-Zee K, Dickmanns A, Fanning E (1992) Intracistronic complementation reveals a new function of SV40 T antigen that co-operates with Rb and p53 binding to stimulate DNA synthesis in quiescent cells. Oncogene 7: 837–847.

Eick D, Hermeking H (1996) Viruses as pacemakers in the evolution of defence mechanisms against cancer. Trends Genet. 12: 4–6.

Ellis RE, Yuan JY, Horvitz HR (1991) Mechanisms and functions of cell death. Annu. Rev. Cell Biol. 7: 663–698.

Evan GI, Littlewood TD (1993) The role of c-*myc* in cell growth. Curr. Opin. Genet. Dev. 3: 44–49.

Evan GI, Wyllie AH, Gilbert CS, Littlewood TD, Land H, Brooks M, Waters CM, Penn LZ, Hancock DC (1992) Induction of apoptosis in fibroblasts by c-myc protein. Cell 69: 119–128.

Fanidi A, Harrington EA, Evan GI (1992) Co-operative interaction between c-myc and bcl-2 proto-oncogenes. Nature 359: 554–556.

Hermeking H, Eick D (1994) Mediation of c-myc-induced apoptosis by p53. Science 265: 2091–3209.

Hermeking H, Wolf DA, Kohlhuber F, Dickmanns A, Billaud M, Fanning E, Eick D (1994) Role of c-*myc* in simian virus 40 large tumor antigen-induced DNA synthesis in quiescent 3T3-L1 mouse fibroblasts. Proc. Natl. Acad. Sci. USA 91: 10412–10416.

Kalderon D, Smith AE (1984) *In vitro* muta-genesis of a putative DNA binding do-main of SV40 large-T. Virology 139: 109–137.

Kohlhuber F, Hermeking H, Graessmann A, Eick D (1995) Induction of apoptosis by the c-myc helix-loop-helix/leucine zipper domain in mouse 3T3-L1 fibroblasts. J. Biol. Chem. 270: 28797–28805.

Tjian R, Fey G, Graessmann A (1978) Biolo-gical activity of purified simian virus 40 T antigen proteins. Proc. Natl. Acad. Sci. USA 75: 1279–1283.

Wyllie AH (1980) Glucocorticoid-induced thymocyte apoptosis is associated with endogenous endonuclease activation. Nature 284: 555–556.

10 Role of the InsP$_3$ Receptor in Intracellular Ca^{2+} Release and Ca^{2+} Entry

R.S. Mathias and H.E. Ives

Content

1 Introduction

The microinjection technique provides an alternative to molecular genetics to study the role of specific proteins that are involved in signal transduction. In this study, we microinjected macromolecular inhibitors of the inositol trisphosphate (InsP$_3$) receptor to examine its role in the mobilization of intracellular calcium (Ca^{2+}$_i$) and extracellular Ca^{2+} entry following activation of the platelet-derived growth factor (PDGF) receptor.

The microinjection techniques provide two major advantages for this type of study. First, it is possible to introduce highly specific macromolecular inhibitors – for example, antibodies. Secondly, in contrast with many molecular ge-

Methods and Tools in Biosciences and Medicine
Microinjection, ed. by J. C. Lacal et al.
© 1999 Birkhäuser Verlag Basel/Switzerland

netic techniques, it is possible to introduce inhibitors at well defined time points. Genetic mutations often must be present throughout the life of the cell, with the potential to cause a variety of unwanted changes in cell function. By using microinjection to analyze the consequences of blocking the action of a specific protein in a signaling pathway, we can gain insight into the role that this protein/pathway plays in generating a particular cellular response.

The cellular response we studied is the biphasic increase in intracellular Ca^{2+} induced by activation of the PDGF receptor, a member of the tyrosine kinase receptor family (Huang et al., 1991). In the initial phase, 1,4,5-InsP$_3$ is formed and is believed to mediate release of intracellular Ca^{2+} stores (Berridge, 1993). In the second phase, Ca^{2+} enters the cytoplasm from outside the cell. The mechanism for this latter process remains poorly defined. The most widely accepted model for regulation of Ca^{2+} entry, termed the "capacitative" model (Putney Jr, 1990), suggests that the signal for Ca^{2+} entry comes from depletion of InsP$_3$-dependent intracellular Ca^{2+} stores.

Although the capacitative model has appeal, not all of the currently available data fits with it. Receptor-mediated Ca^{2+} entry appears to occur in the absence of intracellular Ca^{2+} mobilization in some systems (e.g., see Byron et al., 1992, Stauderman and Pruss, 1989). Similarly, in studies investigating the mechanisms of PDGF-mediated Ca^{2+} entry, this laboratory (Huang et al., 1991) and others (Estacion and Mordan, 1993, Ma et al., 1996) have demonstrated that PDGF-induced Ca^{2+} entry can also occur in the absence of intracellular Ca^{2+} release.

To shed additional light on this question, we used microinjection to introduce heparin (Ghosh et al., 1988) or a monoclonal antibody (Mab) (18A10) to the InsP$_3$ type 1 receptor (InsP$_3$R1) (Miyazaki et al., 1992). These inhibitors of the InsP$_3$R were used to determine the role of this receptor in the regulation of Ca^{2+} entry following activation of the PDGF receptor. We found that Ca^{2+} entry was intact despite blockade of InsP$_3$-mediated release of intracellular Ca^{2+} stores. This suggests that a pathway other than the capacitative entry system is involved in regulation of Ca^{2+} entry by PDGF.

In this chapter, we present the following protocols:
- Microinjection technique for introduction of Ca^{2+}-sensitive fluorescent dyes, fura-2-pentapotassium, and Ca^{2+} green-1-hexapotassium salt.
- Method to determine the effectiveness of heparin, an intracellular InsP$_3$R inhibitor, on Ca^{2+}_i mobilization following photolysis of caged-1,4,5-InsP$_3$.
- Method to determine the role of InsP$_3$R in the regulation of Ca^{2+} entry by PDGF using microinjected InsP$_3$R inhibitors heparin and Mab 18A10.

2 Materials

Chemicals and cell culture

4-bromo A-23187, free acid	Molecular Probes	B-1494
Caged D-myo-Inositol-1,4,5-trisphosphate	Calbiochem	407135
Calcium green-1-hexapotassium salt	Molecular Probes	C-3010
Fetal bovine serum	HyClone Labs	A1115-D
Fura-2-pentapotassium salt	Molecular Probes	F-1200
Geneticin G418	Boehringer Mannheim	1–464–973
Low molecular weight heparin	Sigma	H-5271
Monoclonal antibody (Mab) to IgG2a	Sigma	M-9144
Monoclonal antibody to the InsP₃ receptor type 1 (18A10)	Dr. K Mikoshiba, Univ. of Tokyo	
PDGF-BB	Boehringer Mannheim	1–276–956
Polyclonal antibody to IgG	Sigma	I-5381
apo-Transferrin	Sigma	T-5391

- **Cell line preparation:** CHO-PDGF were stable transfectants of Chinese hamster ovary (CHO) cells containing the cloned PDGF-BB receptor (Fantl et al., 1992).
- **Cell culture growth select medium:** Ham's F-12 medium, 10% (v/v) fetal bovine serum, penicillin (50 units/ml), streptomycin (50 units/ml) and 400 µg/ml G418.
- **Cell culture quiescent medium:** Ham's F-12 medium, bovine serum albumin (0.5 mg/ml), apo-transferrin (5 mg/ml), penicillin (50 units/ml), streptomycin (50 units/ml) and 400 µg/ml G418.

Equipment

- Cell culture incubator
- Microscope circular cover glass (25 mm), Fisher, 12–545–102
- High vacuum grease, Dow Corning
- Temperature-controlled chamber
- Nikon epifluorescence inverted microscope
- Rotating holder for excitation filters (340, 380 and 490 nm)
- Labmaster interface board and IBM style computer /UMANS software
- Xenon arc flashlamp
- Narishige micromanipulator Model MO-188
- Borosilicate capillary tubing/omega dot (0.9 mm OD) FHC 30–31–0
- Flaming/Brown Micropipette Puller (Sutter Inst. Co.) Model P-87

Solutions

- **Assay medium (Solution 1)** (Huang et al., 1987)
    - 140 mM NaCl
    - 5 mM KCl
    - 1 mM Na_2HPO_4
    - 25 mM glucose
    - 25 mM Hepes/NaOH (pH=7.2)
    - 0.5 mg/ml bovine serum albumin with 2 mM or without ("0" medium) added $CaCl_2$
- **Microinjection buffer (Solution 2)** (Bird et al., 1992)
    - 27 mM K_2HPO_4
    - 8 mM NaH_2PO_4
    - 26 mM KH_2PO_4 (pH = 7.3)

3   Methods

3.1   Cell culture

CHO-PDGF cells were grown in cell culture growth selection medium in a humidified atmosphere of 5% CO_2, 95% air at 37°C. Culture medium was changed every 2–3 days until cells were confluent at which time cells were passaged, then plated on glass coverslips in six-well cluster culture dishes in preparation for Ca^{2+}_i measurements.

3.2   Preparation of microinjection capillary tubing

Pipettes were made from borosilicate glass tubing using a Flaming/Brown Micropipette Puller. The pull settings were as follows:
1. Heat = 345, Pull = 0, Vel. = 20, Time = 200
2. Heat = 325, Pull = 0, Vel. = 20, Time = 200
3. Heat = 315, Pull = 95, Vel. = 45, Time = 60

3.3   Single cell Ca^{2+}$_i$ measurements with fura-2

Protocol 1   Single cell Ca^{2+}$_i$ measurements with fura-2

1. Cells were plated at 500 cells/ml on a microscope cover glass and grown in serum-containing medium.

2. Twenty-four h prior to Ca^{2+}$_i$ measurements, the cells were placed in quiescent medium to eliminate any residual PDGF activity in the serum.

3. The coverslips were then placed on an open, centrally-drilled Petri dish (35 × 10 mm) and sealed with high vacuum grease.

4. The Petri dish with the attached cover slip was placed on the temperature-controlled chamber mounted on the microscope.

5. Single cells were microinjected with microinjection buffer (**Solution 2**) containing 5 mM fura-2 pentapotassium salt with and without concentrations of heparin or Mab 18A10 10-fold higher than the final desired cytoplasmic concentration.

6. Single cells were microinjected using glass capillary tubing held in a Narishige micromanipulator. The microinjection pipette was introduced into the cell cytosol near the nucleus at 3 o'clock.

7. Volumes of approximately 5 × 10^{-14} l are microinjected by this method (Graessmann et al., 1980).

   Note:
   The success of the microinjection was determined by the appearance of fura-2 fluorescence limited to the cellular boundary. Two observations that we have noted which confirmed that the cell tolerated the microinjection procedure and is viable for study include (1) lack of cellular vacuolization and (2) a steady state basal ratio as determined by the intensities obtained at 340 and 380 nm (see below).

8. After microinjection, the cells were allowed to equilibrate for 20 min before Ca^{2+}$_i$ measurements were performed. Fura-2 fluorescence was calibrated and measured using a Nikon epifluorescence inverted microscope fitted with a rotating holder for excitation filters (340 and 380 nm) as previously described (Huang et al., 1991).

3.4   Single cell Ca^{2+}$_i$ measurements with Ca^{2+} green

Single cells were prepared as above and microinjected with microinjection buffer (**Solution 2**) containing Ca^{2+} green-1-hexapotassium salt (0.1 mM) and concentrations of caged InsP$_3$ with or without heparin 10-fold higher than the final desired cytoplasmic concentration. The success of the microinjection was determined as for fura-2. The fluorescence signal for the photorelease experiments was measured with excitation light passed through a narrow-band interference filter centered at 490 nM to excite Ca^{2+} green-1 fluorescence. Increases in fluorescence emission collected at 515 nM correspond to increases in the free Ca^{2+}_i concentration (Bird et al., 1992).

3.5 Photolysis of caged compounds

Photolysis of microinjected caged 1,4,5-InsP$_3$ was achieved using a modification of the method of Bird et al. (1992). Light from a continuously burning xenon arc flashlamp was passed through a broad 120 nm-band width filter centered at 350 nm and directed to the specimen for times ranging from 0.5 – 2.0 s. The timing and duration of the UV flashes was controlled by the computer. Immediately following these computer-controlled flashes, Ca^{2+}_i measurements were started.

3.6 Statistics

The data are presented as the mean ± SD.

4   Applications

4.1 Microinjected heparin blocks intracellular Ca^{2+} release, but not Ca^{2+} entry, following PDGF

To illustrate the biphasic Ca^{2+} transients, a single CHO-PDGF cell microinjected with fura-2 was first exposed to PDGF-BB (25 ng/ml) in "0" Ca^{2+} medium, followed by re-addition of 2 mM Ca^{2+} to the medium (Fig. 1A). For the purposes of this study, the first Ca^{2+} transient, measured in "0" Ca^{2+} extracellular medium, was termed "Ca^{2+} release", and the second, measured upon readdition of 2 mM Ca^{2+} to the medium, was termed "Ca^{2+} entry".

To determine the effect of blocking the InsP$_3$R on PDGF-mediated Ca^{2+} release and Ca^{2+} entry, we used microinjection to introduce low molecular weight heparin into single CHO-PDGF cells. Heparin (5 mg/ml injectate) completely blocked Ca^{2+} release from intracellular stores following exposure to PDGF (Fig. 1B). On the other hand, the second phase (Ca^{2+} entry) was unaffected by heparin.

Figure 1

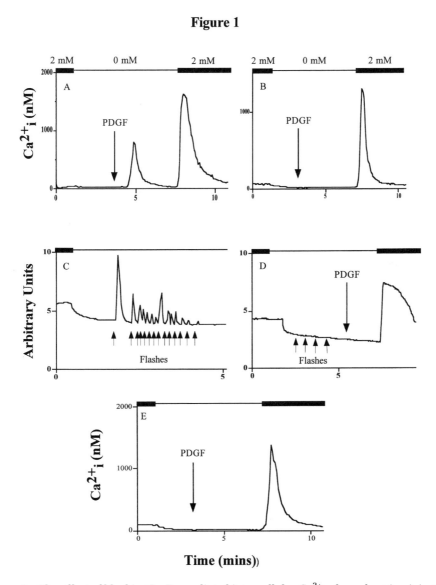

Figure 1 The effect of blocking InsP$_3$-mediated intracellular Ca^{2+} release by microinjected heparin or Mab 18A10 (to the InsP$_3$ Type 1 Receptor) on PDGF-mediated Ca^{2+} entry.

Prior to Ca^{2+} measurements, single CHO-PDGF cells were microinjected with 5 mM fura-2 alone (A), 5 mM fura-2 and 5 mg/ml heparin (B), 0.1 mM Ca^{2+} green and 0.375 mM caged 1,4,5-InsP$_3$ (C), 0.1 mM Ca^{2+} green, 0.375 mM caged 1,4,5-InsP$_3$ and 10 mg/ml heparin (D), or 5 mM fura-2 and Mab 18A10 (0.75 mg/ml) (E). 20 to 30 min. later, Ca^{2+}$_i$ transients (nM in A, B and E, and in dimensionless fluorescence units in C, D) were measured. In A, B and E, PDGF (25 ng/ml) was added in "0" Ca^{2+} medium followed by the readdition of 2 mM Ca^{2+} medium. In C, each spike represents a single Ca^{2+}$_i$ transient in response to single flashes of UV light (each *arrow* represents a single flash of 0.5–2 s duration) in "0" Ca^{2+} medium. In D, after microinjection of heparin, Ca^{2+}$_i$ failed to increase despite multiple single flashes of UV light (0.5 to 4 s duration) and the addition of PDGF (25 ng/ml) in "0" Ca^{2+} medium, followed by the readdition of 2 mM Ca^{2+} medium. Medium Ca^{2+} concentration was varied between "0" mM (–) and 2 mM Ca^{2+} (▬). Tracings shown are representative of at least 5–16 experiments.

To demonstrate heparin's ability to specifically inhibit InsP$_3$-mediated intracellular Ca^{2+}$_i$ release, we microinjected Ca^{2+} green (0.1 mM) and caged 1,4,5-InsP$_3$ (0.375 mM) into CHO-PDGF cells in the absence or presence of heparin (5–20 mg/ml injectate concentration). Photorelease of InsP$_3$ was achieved with multiple flashes of UV light of 0.5–2 s. duration. Each flash produced a single Ca^{2+} spike from cells in "0" Ca^{2+} medium (Fig. 1C). Heparin (10 mg/ml) blocked intracellular Ca^{2+} release in "0" Ca^{2+} medium following both flash photolysis of caged 1,4,5-InsP$_3$ or PDGF. As seen in Figure 1B, Ca^{2+} entry was again unaffected by microinjected heparin (Fig. 1D).

4.2   Competitive inhibition of PDGF-dependent intracellular Ca^{2+} release by microinjected heparin over a wide concentration range

In further consideration of the competitive inhibitory nature of heparin on InsP$_3$R-mediated Ca^{2+} release (Ghosh et al., 1988), we microinjected various concentrations of heparin (Table 1) prior to stimulation by PDGF (25 ng/ml). As the microinjected injectate concentration heparin was decreased from 1.0 mg/ml to 0.01 mg/ml, the percentage of cells that produced a Ca^{2+} response to PDGF in "0" Ca^{2+} medium increased in a dose-dependent manner. Half-maximal inhibition of Ca^{2+} release was between 0.1 mg/ml and 0.3 mg/ml heparin and nearly complete blockade was observed when heparin injectate concentration was greater than 1.0 mg/ml.

Table 1   Dose response of heparin blockade of PDGF-induced release of intracellular Ca^{2+} stores.

Low molecular weight heparin was microinjected into CHO cells expressing the PDGF receptor. 20 to 30 min later, the intracellular Ca^{2+} response was measured in following "0" Ca^{2+} medium exposure to PDGF (25 ng/ml). The data are expressed as percent of cells responding to the ligand in the presence of the indicated concentration of microinjected heparin. The number of single experiments at each concentration are shown in parentheses.

Heparin (mg/ml)	PDGF (% Responsive)	n
0.01	100	(4)
0.1	62	(8)
0.3	33	(6)
0.5	20	(10)
1.0	11	(9)
5.0	7	(27)
10.0	11	(9)

4.3   InsP$_3$ R1 Mab blocks release of intracellular Ca^{2+} stores, but fails to block Ca^{2+} entry following PDGF

Since the specific inhibitory effects of heparin on InsP$_3$R-mediated Ca^{2+} release may not be complete (Guillemette et al., 1989), we used microinjection to examine the effects of a microinjected Mab to the mouse InsP$_3$R1 (18A10) (Miyazaki et al., 1992) on PDGF-mediated Ca^{2+} release and Ca^{2+} entry. Microinjected Mab 18A10 (injectate concentrations of 0.25 to 0.75 mg/ml) completely blocked PDGF-induced release of intracellular Ca^{2+} stores but did not block subsequent Ca^{2+} entry (Fig. 1E). Microinjection of nonspecific antibodies (IgG2α or a polyclonal antibody to IgG) had no effect on PDGF-mediated Ca^{2+} release or Ca^{2+} entry (data not shown). Taken together, these data suggest that PDGF-mediated Ca^{2+} entry can occur despite blockade of InsP$_3$R-dependent intracellular Ca^{2+} release by heparin or Mab 18A10 to the InsP$_3$R1.

5   Remarks and Conclusions

In this study, we used microinjection to introduce macromolecules that interfere with the InsP$_3$-InsP$_3$R pathway in order to examine its role in the regulation of PDGF-mediated Ca^{2+} entry. In the "capacitative" model of Ca^{2+} entry, depletion of InsP$_3$-dependent intracellular stores is proposed to generate a signal that opens Ca^{2+} channels on the plasma membrane, causing Ca^{2+} influx from outside the cell (Putney Jr, 1990). A major weakness of the capacitative model is that the signal necessary to couple the state of the intracellular stores to the Ca^{2+} entry channel has never been clearly defined.

As discussed in the Introduction, Ca^{2+} entry has been demonstrated to occur without prior depletion of intracellular Ca^{2+} stores (i.e., by a non-capacitative mechanism) in a variety of systems (Stauderman and Pruss, 1989, Dawra et al., 1993). Earlier work in this laboratory showed that microinjection of heparin (4 mg/ml injectate) completely blocked PDGF-mediated release of intracellular Ca^{2+} stores, but did not effect subsequent Ca^{2+} entry in single vascular smooth muscle cells (Huang et al., 1991). We now extend these findings by showing that microinjection of CHO cells with heparin or anti-InsP$_3$ receptor antibodies (Mab 18A10) blocks PDGF-mediated intracellular Ca^{2+} release but fails to block Ca^{2+} entry.

Our finding that microinjected heparin fails to block Ca^{2+} entry is consistent with recent work by Ma et al. (1996) showing that PDGF was capable of causing Ca^{2+} entry currents in excised patches of mesangial cells even in the presence of heparin. Furthermore, previous work from our laboratory showed that ongoing PDGF receptor occupancy, and not depletion of Ca^{2+} stores, was essential to the generation of Ca^{2+} entry in vascular smooth muscle cells (Huang et al., 1991). The signal that is produced in cell systems capable of non-capacitative Ca^{2+} entry has not been characterized. Previous work with anti-PIP_2 antibodies (Huang et al., 1991) and PDGF receptor mutants (unpublished) indicate that this signal derives from PLCγ-mediated breakdown of PIP_2. Potential alternatives to $InsP_3$ include other inositol phosphates such as $InsP_4$, or lipid products of PIP_2 breakdown, which may affect Ca^{2+} entry via activation of protein kinase C isoforms.

In summary, we find that release of intracellular Ca^{2+} stores by PDGF is blocked by microinjected heparin and Mab to the $InsP_3R1$ 18A10, showing that PDGF-mediated Ca^{2+} entry can occur without activation of the capacitative system. The technique of microinjection was critical in obtaining these results, and can be a powerful tool in answering questions about cellular signal transduction.

Acknowledgements

This work was supported by National Institutes of Health Grant HL – 41210 (H.E.I.), a Grant in Aid from the American Heart Association (H.E.I.), and the University of California, San Francisco School of Medicine REAC Blair Fund (R.S.M.).

References

Berridge MJ (1993) Inositol trisphosphate and calcium signalling. Nature 361: 315–325.

Bird GSJ, Obie JF, Putney Jr. JW (1992) Sustained Ca^{2+} signaling in mouse lacrimal acinar cells due to photolysis of "caged glycerophosphryl-myo-inositol 4,5-bisphosphate. J Biol Chem 267: 17722–17725.

Byron KL, Babnigg G, Villereal ML (1992) Bradykinin-induced Ca^{2+} entry, release, and refilling of intracellular Ca^{2+} stores. J Biol Chem 267: 108–118.

Dawra R, Saluja A, Runzi M et al. (1993) Inositol trisphosphate-independent agonist-stimulated calcium influx in rat pancreatic acinar cells. J Biol Chem 268: 20237–30242.

Estacion M, Mordan L (1993) Competence induction by PDGF requires sustained calcium influx by a mechanism distinct from storage-dependent calcium influx. Cell Calcium 14: 439–454.

Fantl WJ, Escobedo JA, Martin GA et al. (1992) Distinct phosphotyrosine on a growth factor receptor binds to specific molecules that mediate different signaling pathways. Cell 69: 413–423.

Ghosh TK, Eis PS, Mullaney JM et al. (1988) Competitive, reversible, and potent antagonism of inositol 1,4,5-trisphosphate-activated calcium release by heparin. J Biol Chem 263: 11075–11079.

Graessmann A, Graessmann M, Mueller C (1980) Microinjection of early SV40 DNA fragments and T antigen. Methods Enzymol 65: 816–825.

Guillemette G, Lamontagne S, Boulay G et al. (1989) Differential effects of heparin on inositol 1,4,5-trisphosphate binding, metabolism, and calcium release activity in the bovine adrenal cortex. Mol Pharm 35: 339–344.

Huang CL, Cogan MG, Cragoe EJ et al. (1987) Thrombin activation of the Na^+/H^+ exchanger in vascular smooth muscle cells. J Biol Chem 262: 14134–14140.

Huang CL, Takenawa T, Ives HE (1991) Platelet-derived growth factor-mediated Ca^{2+} entry is blocked by antibodies to phosphatidylinositol 4,5-bisphosphate but does not involve heparin-sensitive inositol 1,4,5-trisphosphate receptors. J Biol Chem 266: 4045–4048.

Ma H, Matsunaga H, Li B et al. (1996) Ca^{2+} channel activation by platelet-derived growth factor-induced tyrosine phosphorylation and ras guanine trisphosphate-binding proteins in rat glomerular mesangial cells. J Clin Invest 97: 2332–2341.

Miyazaki S, Yuzaki M, Nakada K et al. (1992) Block of Ca^{2+} wave and Ca^{2+} oscillation by antibody to the inositol 1,4,5-trisphosphate receptor in fertilized hamster eggs. Science 257: 251–255.

Putney Jr JW (1990) Capacitative calcium entry revisited. Cell Calcium 11: 611–624.

Stauderman KA, Pruss RM (1989) Dissociation of Ca^{2+} entry and Ca^{2+} mobilization responses to angiotensin II in bovine adrenal chromaffin cells. J Biol Chem 264: 18349–18355.

11 Microinjection of *Xenopus laevis* Oocytes: A Model System

J.C. Lacal

Contents

1   Introduction

Oocytes of the African frog *Xenopus laevis* have been used extensively as a system to study the biological activity of many different molecules. It is an excellent system to study regulation of transcription, translation and secretion, as well as the metabolic alterations associated with signaling cascades. The large size of fully grown oocytes (over 1.2 mm in diameter) and their easy manipulation and evaluation of results, has made this system an excellent one to study the potential role of any type of molecule in either mitogenic signaling or a variety of metabolic pathways. *Xenopus* oocytes have been specially useful in the discovery and identification of enzymes involved in the regulation of the cell cycle, and have been used extensively for the analysis of potential mitogenic signals, expression and characterization of many kinds of receptors, identification of potential inhibitors of signaling molecules, and more recently, also as a model system for the identification of apoptosis-related molecules. This research is based on the fact that the biological effects of these signals can be easily recognized by monitoring either the breakdown of the germinal vesicle (GVBD), biochemical changes within the oocyte, or DNA degradation in apoptotic nuclei.

Methods and Tools in Biosciences and Medicine
Microinjection, ed. by J. C. Lacal et al.
© 1999 Birkhäuser Verlag Basel/Switzerland

Stage VI oocytes from *Xenopus laevis* are extremely large cells (1.2–1.4 mm in diameter) arrested at the G2/M border of the first meiotic prophase. They can be easily injected, and their size allows for biochemical characterization of the injected cells. An example of stage VI oocytes is shown in Figure 1A. After stimulation, the oocytes reenter the cell cycle and a burst of intracellular phosphorylation and biochemical alterations occur (for extensive review, see Sagata, 1996; Page and Orr-Weaver 1997). A few hours after hormone treatment, the oocyte nucleus (germinal vesicle), which is located near the center of the oocyte, migrates towards the animal hemisphere surface and begins the process of dissolution. The arrival of the nucleus at the cortex displaces cellular pigment, producing a white circular spot. This white spot is the first visible indication that oocyte maturation is proceeding. After dissolution of the nuclear membrane, a process known as germinal vesicle break-down (GVBD), the condensed chromosomes align on the first metaphase spindle where they remain until the mature egg is fertilized. An example of oocytes after GVBD has taken place is shown in Figure 1B.

Since GVBD is the easiest event to score, it has frequently been used as the criterion of maturation. Recent development on the knowledge of the signaling cascades, along with the identification of the molecules involved in these processes has made it possible to use other strategies for the study of oocytes maturation. These biochemical analyses are far more informative and precise than reporting GVBD alone and will constitute most of the work in the future in this field. Some of these procedures, such as activation of the MPF complex or the MAPK pathway, are fully described in Chapter 12.

GVBD, an assay for oocytes maturation

Oocyte maturation has been studied in a variety of vertebrate and invertebrate organisms, but the process has been investigated most extensively in amphibians. As indicated above, *Xenopus* oocytes are arrested at the G2/M border of the first meiotic prophase. The resumption of meiosis *in vivo* is brought about by the action of a gonadotropic hormone which acts on ovarian follicle cells, causing them to produce progesterone which acts directly on the oocyte to initiate the process of oocyte maturation. Similarly, progesterone induces maturation *in vitro* in oocytes dissected from their ovarian follicles (Wasserman and Smith, 1978). A few hours after hormone treatment, the germinal vesicle situated near the center of the oocyte, starts to migrate towards the animal hemisphere surface and dissolve.

An oocyte is not mature until it has progressed to the second meiotic metaphase and can be activated. However, since GVBD is the easiest event to score, it frequently has been used as the major if not the only criterion that maturation is underway. An example of normal GVBD in oocytes induced to mature with progesterone is shown in Figure 1B. However, in oocytes from some females, and in oocytes treated or microinjected with certain substances, the nucleus will rise to the surface, displacing the pigment or altering its distribution in the animal pole, without a real GVBD. This occurs after treatment with cyto-

A

B

C

D

Figure 1   An example of fully grown, stage VI.

Xenopus laevis oocytes is shown in (A). These oocytes have diameters over 1.2 mm with a clear division of a pigmented animal half (color can be from light to dark brown) and a unpigmented vegetal half (actually pale yellow). After progesterone treatment, the nucleus of the oocyte (germinal vesicle) which is near the center of the oocyte, moves to the surface of the animal hemisphere and dissolves (germinal vesicle breakdown, GVBD). A white spot is readily observed as a consequence (B). In addition to progesterone, a large number of compounds can induce the metabolic responses associated to GVBD. As shown in this figure, phorbol esters (C) and high levels of calcium (D) induce GVBD and typical changes on the pigmentation of the oocytes which makes scoring of GVBD more difficult to evaluate.

chalasin or colchicine (Coleman et al., 1981), or after microinjection with a cDNA of protein kinase C without the regulatory domain (Muramatsu et al., 1989), or rap 2 proteins (Campa et al., 1993). Migration *per se*, or pigment alterations are not indicative of maturation. On the other hand, several studies reported that GVBD occurs in response to stimulus, such as phorbol esters and salts, but the oocytes also exhibit obvious degenerative changes which precede the onset of GVBD (Figs 1C and D). The physiological significance of GVBD in these oocytes is difficult to evaluate.

The timing of GVBD can vary considerably in oocytes from different females. This can result in part from differing environmental conditions under which animals are maintained in different laboratories. In addition, a variety of diverse media have been used to culture oocytes and the time at which GVBD occurs after agonist treatment in oocytes can vary as much as two-fold in different media. Finally, the injection of gonadotrophins into females, either to induce ovulation (human chorionic gonadotropins) or to improve the synchrony of response to progesterone (pregnant mare serum gonadotropin) can dramatically alter the time of GVBD. It has been reported that in females injected with gonadotropins a few days before oocyte extraction, GVBD occurred within 2 h of progesterone exposure; meanwhile in untreated females the progesterone-induced GVBD proceed out normally at 6–8 h after progesterone exposure.

In addition to progesterone which is the physiological inducer of maturation, several other steroid hormones as well as a large number of seemingly diverse drugs and chemicals are reported to induce oocyte maturation. A partial list of these various agonists is compiled in Table 1. It should be noted that, in some cases, the only assay for maturation used is GVBD. These results should be confirmed by a more solid evidence, using the molecular markers identified in the last few years, some of which are described in detail in Chapter 12.

Table 1   Inducers of GVBD in *Xenopus* oocytes.

Agent	Reference
Steroids	
Progesterone, testosterone, cortisol, pregnenolone	Balieu et al., 1978
Hormones	
Insulin, insulin-like growth factor	Maller and Koontz, 1981
Proteins	
MPF	Masui and Markert, 1971
R subunit of PKA	see Smith, 1989
cyclin A and B	see Smith, 1989
c-mos	Sagata et al., 1989
ras-p21	Birchmeier et al., 1985
Phospholipases C, D and A2	Carnero and Lacal, 1993
PKC ζ	Carnero and Lacal, 1995
calmodulin	see Smith, 1989
Ions	
Ca^{2+}, Zn^{2+}, Co^{2+}, Ba^{2+}, Mg^{2+}	Cicirelli and Smith, 1987
Drugs	
propranolol, alprenolol	Baulieu et al., 1978
A23187 plus Mg^{2+}, Lanthanum verapamil	Baulieu et al., 1978
Tetracaine, trifluoperazine	Hollinger and Alvarez, 1982
Lipids	
PA, AA, DAG, lysoderivates	Carnero and Lacal, 1993

2   Materials

Equipment

A large number of equipment that is commercially availabe provides the required set-up for preparation of micropipets and performing microinjection successfully. Since there is great flexibility in the choice of equipment, only those that the author is more familiar with are described here:

- For microinjections, almost any type of *stereo microscope* can be used. I have used either C. Zeiss GSZ, C. Zeiss Stemi SR, or Nikon SMZ-2B stereo microscopes. An appropriate source of light such as KL 1500-Z (C. Zeiss) or equivalent should be used to avoid overheating of the oocytes.
- *Micromanipulators:* many different set ups are commercially available. I have used either a vertical micromanipulator from Narishige, or the MKI micromanipulator (Singer Instruments).
- *Pressure system:* several pumps are available on the market. I have used the Inject+Matic air pump (A. Gabay, Geneve, Switzerland) which does not require pressurized tanks and is triggered by a foot pedal, and the PLI100 Pressure Source from Medical Systems Corp. connected to a N2 tank for pressure. The pressure and time of injection will be experimentally determined by calibration of each capillary using some oocytes that will be discarded afterwards.

Solutions

Oocyte incubation buffers

The following buffers are recommended to be used for the dissection and conservation of oocytes for short-term experiments (2–3 days) at room temperature:

- **Ringer's (Solution 1)**
    - 100mM NaCl
    - 1.8 mM KCl
    - 2 mM $MgCl_2$
    - 1 mM $CaCl_2$
    - 4 mM $NaHCO_3$ pH 7.8
    - This buffer is stable for months at room temperature
- **OR2 (Solution 2)**
    - 82.5 mM NaCl
    - 2.5 mM KCl
    - 1 mM $CaCl_2$
    - 1 mM $MgCl_2$
    - 1 mM Na_2HPO_4
    - 5 mM HEPES pH 7.8
    - OR2 is prepared in two 10x stock solutions: solution A (NaCl, KCl, Na_2HPO_4, Hepes and NaOH to pH 7.8) and solution B ($CaCl_2$, $MgCl_2$). Both solutions ar mixed immediately before use. It is important to check that the pH of the resulting solution is 7.8.

Alternatively, the following buffers can be used with good results:

- **HEPES (Solution 3)**
    83 mM NaCl
    1 mM $MgCl_2$
    0.5 mM $CaCl_2$
    1 mM KCl
    10 mM HEPES pH 7.9
- **Barth's (Solution 4)**
    88 mM NaCl
    2.4 mM $NaHCO_3$
    1 mM KCl
    0.82 mM $MgSO_4$
    0.74 mM $CaCl_2$
    0.33 mM $Ca(NO_3)_2$
    10 mM Tris-ClH pH 7.6
- **De Boer's (Solution 5)**
    110 mM NaCl
    1.3 mM KCl
    4.4 mM $CaCl_2$
    $NaHCO_3$ to pH 7.2

Note:
All buffers can be supplied with antibiotics (penicillin 50 units/ml, streptomycin 50 mg/ml, amphotericin B 125 ng/ml) and BSA (1 mg/ml).

Microinjection buffers
Aqueous buffers can be used for microinjection. An excessive concentration of ions should be avoided since high concentrations of some ions are able to induce GVBD themselves. Buffers frequently used to keep the activity of proteins or other compounds are good candidates for microinjection studies. It is critical that pH values are kept between 6.8 and 7.2. As examples, 20 mM MES pH 7.0, or 50 mM Tris-HCl pH 7.0 can be used with excellent results. Microinjection buffers can be supplemented with fatty-acid-free bovine serum albumin at concentrations of 0.5–1 mg/ml.

3   Methods

3.1   Preparation of the oocytes

Xenopus laevis females can be obtained from many companies in Europe, Africa or the USA. In order to obtain good quality oocytes, the animals should be maintained under appropriate conditions. We have noticed that noisy rooms should be avoided since this condition will negatively affect the quality of the oocytes. Also, use of the oocytes soon after arrival to destination should be avoided. A minimum of 2 weeks would be appropriate to reduce stress.

Keep the animals in tanks or aquaria with de-chlorinated water (tap water aged for several days) at 23°C. Best conditions are achieved with circulating water that can be produced with an electric immersion pump at the bottom. This will create the appropriate aeration of the water. Animals should be fed twice a week, and the water changed entirely the following day. We have obtained excellent results with commercial granulated food. Oocytes can be used from untreated females or females treated with human chorionic gonadotropin (hCG). Treatment with hCG hormone should be used with care, since it stimulates steroids production by the gonads and facilitates maturation of the oocytes. By affecting the metabolic activity of the treated oocytes, hCG may alter the response to other stimuli, and therefore this fact must be taken into consideration when investigating signaling pathways. The hormone can be purchased from several companies (Sigma, Organon, etc). Treatment of the animals is carried out by injection into the dorsal lymph sac. For treatment, 100–200 U of hCG are injected 2 weeks prior to oocytes collection. Regardless of the treatment, oocytes are selected according to the following protocol.

Protocol 1   Oocyte selection

1. Animals must be anesthetized prior to operation. This can be achieved by either immersion for about 30 min into a 0.2% solution of ethylamino benzoate or a solution of 0.1% tricane-methane sulfonate, or simply by a cold shock in ice for at least 30 min.

2. Once anesthetized, the animal is placed on its back on a clean surface covered with aluminum foil. If anesthetized by a cold shock, ice should be placed underneath the aluminum foil to keep the animal unsensitive. Wipe the abdomen with 70% ethanol.

3. Make a small lateral incision about 1 cm in length through the skin and body wall just above the leg using sterilized tools (scalpel, forceps, scissors). Pull out the ovary through this ventral incision with a watchmaker's forceps and cut as many pieces as required with sterilized scissors. You should realize that adult *Xenopus* females have thousands of oocytes at all six stages of oocytes growth. For most of the experiments, only a few hundred of fully grown oocytes (stage VI) will be required. Therefore, the same animal can be used several times.

4. After operation, wounds in both the body wall and the skin are sutured separately with several stitches of sterile catgut and the animal is left to recover on a water bath at room temperature. Females can be operated on several times before being sacrificed under anesthesia.

5. Each fully mature *Xenopus laevis* female carries several thousands of oocytes of >1 mm in diameter. Injections are performed with stage VI oocytes which are characterized by their size and appearance: over 1.2 mm in diameter, unpigmented equatorial band and a dark animal hemisphere (see Fig. 1A).

6. Oocytes can be microinjected after selection with no further treatment. However, they can be injected more easily after defolliculation. This can be achieved by incubating small pieces of the ovaries cut with scissors in Ringer's medium containing 0.2% collagenase (type I Sigma) at 18–22°C under smooth agitation. Usually, after 2–3 h of incubation, individual oocytes start to be freed from the follicular tissue. At this time, they should be washed exhaustively and transferred to flat dishes with fresh medium and kept at least several hours before injections.

   Note:
   Since collagenase may damage oocytes and may interfere with their maturation process if used a few hours after treatment, I usually do not use collagenase-treated oocytes. If collagenase has to be used, I recommend to use the oocytes after overnight recovery. Oocytes can be also stored overnight at room temperature to let them fully recover and to avoid use of those damaged during treatment. Oocytes should always be transferred by wide-mouthed pipets to avoid stress and lysis.

7. Alternatively, oocytes can be manually dissected by pulling individual oocytes away from the follicles. For this, a small piece of the ovary containing a few hundred follicles is cut with scissors with a microbiological sterile loop while holding the piece of the ovary with a watchmaker's forceps. Grasp gently to free individual oocytes and select those with an intact appearance.

8. Oocytes are then selected according to their size and their appearance. Selected oocytes are then transferred carefully to a fresh dish with a wide-mouthed pipet, avoiding as much as possible taking buffer if turbid, an indication of oocyte lysis or leakiness. Only stage VI oocytes are selected with a diameter of at least 1.2 mm, eliminating those that have turned white. Buffers recommended for handling oocytes can be supplemented prior to use with antibiotics to avoid contaminations (penicillin 50 units/ml, streptomycin 50 mg/ml, amphotericin B 125 ng/ml) and fatty-acid free BSA (1 mg/ml) to avoid that the oocytes stick to each other. Gentle vibration of the dish usually is sufficient to turn the oocytes to the appropriate position for microinjections. Otherwise, a gentle buffer purge from the pipet toward the oocytes will have the same effect.

3.2   Preparation of micropipets

Preparation of the injection capillaries is one of the critical stages of the technique. Thus, special care should be taken in this process. Many types of commercial equipment are available which are suitable for fabrication of injection pipets. These consist of an electrically heated solenoid and a pulling device for the glass capillaries. Available pullers must be evaluated regarding the glass used, the diameter of the tip, whether an open or closed tip is generated, and the shape of the capillary. I routinely use the Narishige PP-83 puller, but others also provide excellent results. We prepare micropipets according to the following protocol.

Protocol 2   Preparation of micropipets

1. *Glass capillaries*: borosilicate glass capillaries (Kimble, 1.0 mm OD and 0.7 mm ID) or borosilicate glass capillaries (Clark GC120F-10, 1.2 mm OD; 0.69 mm ID) give excellent results.

2. *Pulling procedure:* The glass pipet is placed through the solenoid clipped by two clamps and heated under stress until the pulling force pulls out the two pieces. The capillary must not touch the heating filament.

3. *Adjust heaters*: number 1 heater: 15; number 2 heater: 13. Ensure that the lowest pulling tension obtainable is applied.

4. Switch on the heaters.

5. Capillaries are drawn with tip diameters of around 0.5–1 µm. Check capillaries under the microscope for their shape. They may be closed as they come out from the puller. To check whether the tip is closed, immerse it into a clean tube with ethanol and inject pressurized air with a syringe connected to a pipet holder. Fine bubbles will come out of the tip if it is open.

6. *Opening of the pipet*: the tip of the pipet should be open after pulling and should have an external diameter of 10–20 µm. Best results are achieved if the tip is polished once pulled using a special ground mill with a 45° opening (Narishige EG-4). However, breaking the tip with watchmaker's forceps can be sufficient to generate efficient pipets of about 10–20 µm tips.

   Note:
   Store capillaries in a dust-free and dry environment until use. A plastic plate containing a small piece of artist's clay will be an excellent way to hold them safely.

7. *Filling capillaries with sample*: pipets are loaded by sucking about 1–3 μl of the experimental solution (proteins, enzymes, lipid metabolites, etc.). In order to avoid capillary clogging the sample is centrifuged 10 min. at 10000 rpm. A drop (2–5 μl) of the solution is placed on a piece of parafilm under the binocular, and the tip of the pipet introduced until it reaches the bottom of the drop. This can be achieved directly with the Medical System Corp. microinjector unit and the Inject+Matic system or connecting a vacuum pump to the Eppendorf microinjection apparatus. Aspiration is controlled by keeping the meniscus within the visual range of the binocular.

8. *Capillary calibration*: for proper estimation of the injected amounts, I use a home-made, graduated lens carrying a grid divided into millimeters and 1/10 millimeters. When the meniscus of the capillary is focused, the grid indicates the amount by direct comparison to the scale. Since the capillaries used have an internal diameter of approximately 0.7 mm, marks on the scale are equivalent to 38 nl of sample per l/10th millimeter.

9. *Injections*: with a defined pressure, inject and visualize the volume injected following the length of the graduated scale. Start with a low pressure and modify it accordingly to the desired volume. If properly adjusted, only one calibration will be needed for each pipet.

3.3 Injection procedure

Injection of the oocytes is the final stage. It requires not only skill and experience, but systematic movements to increase speed and accuracy. Also the most confortable position should be accomplished regarding the binocular and the stage. Injections are controlled by a microinjector pump which provides the required pressure. There are different types of pumps, requiring or not pressurized tanks. The pump must have a foot switch to easily perform injections. The pipets are controlled by a micromanipulator which can be either vertical or tilted at approximately 45°. The following protocol is used in our laboratory.

Protocol 3   Injection procedure

1. Oocytes (25–50) are placed in the center of a plastic dish with a plastic grid glued to it (a drop of chloroform would be an excellent glue). The dish should be filled with buffer.

2. The needle should penetrate into the oocyte by a fifth of its diameter. Erroneous consideration of GVBD due to the appearance of a white spot in the animal hemisphere may be a consequence of the damage infringed by injections. To avoid this problem, it is best to inject the oocytes in the interfaces. A quick movement of the hand, combined with the foot pedal, will execute the injection. At the same time the oocyte is moved to a distant part of the dish where it is released by the surface tension when completely pulling the pipet out of the buffer. This will also facilitate the identification of those oocytes injected from those to be injected.

3. *Injection*: with a defined pressure, inject and visualize the volume injected following the length of the graduated scale. Modify pressure accordingly to the desired volume. Usually only one calibration is needed for each pipet. Cytoplasmic injections are performed with volumes of less than 50 nl per oocyte. This amount accounts for 1/10th of the total cytoplasmic volume, estimated in 500 nl for stage VI oocytes. Injection of a larger volume can damage the oocyte.

Once microinjection has been completed, oocytes are transferred to a multiwell dish containing fresh medium at a rate of 200 μl/oocyte. Treatments of oocytes with hormones and other substances can be performed at this stage. Results are analyzed at different times after injection depending on the experimental procedures.

Acknowledgements

This publication has been possible thanks to the specific support from DGICYT (Grant #PB94–0009) and FIS (Grant #96/2136).

References

Baulieu EE, Godeau F, Schorderet M, Schorderet-Slatkine S (1978) Steroid-induced meiotic division in *Xenopus laevis* oocytes: surface and calcium. *Nature* 275, 593–598.

Birchmeyer C, Broek D, Wigler M (1985) Ras proteins can induce meiosis in *Xenopus* oocytes. *Cell* 43: 615–621.

Campa MJ, Farrell FX, Lapetina EG, Chang KJ (1993) Microinjection of Rap2B protein or RNA induces rearrangement of pigment granules in *Xenopus* oocytes.- *Biochem J.* 292: 231–236.

Carnero A, Jimenez B, Lacal JC (1994) Progesterone but not Ras requires MPF for *in vivo* activation of MAP K and S6 KII. MAP K is an essential conexion point of both dignaling pathways. *J. Cell. Biochem.* 55: 465–476.

Carnero A, Lacal JC (1993) Phospholipase induced maturation of *Xenopus laevis* oocytes. Mitogenic activity of generated metabolites. *J. Cell. Biochem.* 52: 440–448.

Carnero A, Liyanage M, Stabel S, Lacal JC (1995) Evidence for different signalling pathways of PKCζ and ras-p21 in *Xenopus* oocytes. *Oncogene* 11: 1541–1547.

Cicirelli MF, Smith LD (1987) Do calcium and calmodulin trigger maturation in amphibian oocytes? *Devl. Biol.* 121: 48–57.

Coleman A, Morser J, Lane C, Besley J, Mylie C, Velle G (1981) Fate of secretory proteins trapped in oocytes of *Xenopus laevis* by disruption of the cytoskeleton or by imbalanced subunit synthesis. *J. Cell. Biol.* 91: 770–780.

Daar I, Nebreda AR, Yew N, Sass P, Paules R, Santos E, Wigler M, Vande Woude GF (1991) The ras oncoprotein and M-phase activity. *Science* 253: 74–76.

Dunphy GW, Brizuela L, Beach D, Newport J (1988) The *Xenopus* cdc2 protein is a component of MPF, a cytoplasmic regulator of mitosis. *Cell* 54: 423–431.

Erikson RL (1991) Structure, expression, and regulation of protein kinases involved in the phosphorylation of ribosomal protein S6. *J. Biol. Chem.* 266: 6007–6010.

Fabian JR, Morrison DK, Daar IO (1993) Requirement for Raf and MAP kinase function during the meiotic maturation of *Xenopus* oocytes. *J. Cell Biol.* 122: 645–52.

Gotoh Y, Moriyama K, Matsuda S, Okumura E, Kishimoto T, Kawasaki H, Suzuki K, Yahara I, Sakai H, Nishida E (1991) *Xenopus* M-phase MAP kinase: isolation of its cDNA and activation by MPF; *EMBO. J.* 10: 2661–2668.

Hollinger TG, Alvarez IM (1982) Trifluoperazine-induced meiotic maturation in *Xenopus laevis*. *J. Exp. Zool.* 224: 461–464.

Jacobs T (1992) Control of the cell cycle. *Dev. Biol.* 153: 1–15.

Maller JL, Koontz JW (1981) A study of the induction of cell division in amphibian oocytes by insulin. *Dev. Biol.* 85: 309–316.

Masui Y, Markert CL (1971) Cytoplasmic control of nuclear behavior during meiotic maturation of frog oocytes. *J. Exp. Zool.* 177: 129–146.

Muramatsu MA, Kaibuchi K, Arai KI (1989) Protein kinase C cDNA without the regulatory domain is active after transfection *in vivo* in the absence of phorbol ester. *Mol. Cell. Biol.* 9: 831–836.

Murray AW, Kirschner MW (1989) Cyclin synthesis drives the early embryonic cell cycle. *Nature* 339: 275–280.

Murray AW, Solomon MJ, Kirschner MW (1989) The role of cyclin synthesis and degradation in the control of maturation promoting factor activity. *Nature* 339: 280–286.

Page AW, Orr-Weaver TL (1997) Stopping and starting the meiotic cell cycle. *Curr. Op. Genet. Develop.* 7: 23–31.

Sagata N (1996) Meiotic metaphase arrest in animal oocytes: its mechanism and biological significance. *Trends Cell Biol.* 6: 22–28.

Sagata N, Daar I, Oskarsson M, Showalter SD, Vande Woude GF (1989) The product of the mos proto-oncogene as a candidate "initiator" for oocyte maturation. *Science* 245: 643–646.

Smith LD (1989) The induction of oocyte maturation: transmembrane signaling events and regulation of the cell cycle. *Development* 107: 685–699.

Wasserman WJ, Smith LD (1978) The cyclic behavior of a cytoplasmic factor controlling nuclear membrane breakdown. *J. Cell. Biol.* 78: 15–22.

Yew N, Mellini ML, Vande Woude GF (1992) Meiotic initiation by the mos protein in *Xenopus*. *Nature* 355: 649–652.

Xenopus laevis Oocytes as a Model for Studying the Activation of Intracellular Kinases

A. Carnero and J.C. Lacal

Contents

1   Introduction: Induction of cell cycle progression in *Xenopus* oocytes

As indicated in the previous chapter, *Xenopus laevis* oocytes are arrested at the G2/M border of the first meiotic prophase. After progesterone treatment, a complex process of intracellular signals is activated which ends in the maturation of the oocytes, a process reported by the Breakdown of the Germinal Vesicle (GVBD). Among the critical signals turned on, activation of MPF from inactive stores takes place prior to observance of GVBD. This activation is necessary and sufficient to induce maturation. The injection of active MPF induces

Methods and Tools in Biosciences and Medicine
Microinjection, ed. by J. C. Lacal et al.
© 1999 Birkhäuser Verlag Basel/Switzerland

precocious GVBD (Gotoh et al., 1991) and the microinjection of p13^{suc} protein that specifically binds to cdc2 protein (the kinase component of MPF complex) inhibits its kinase activity blocking the progesterone-induced maturation (Dunphy and Newport, 1989). There are two peaks of MPF activity, the first coincides approximately with metaphase I and needs synthesis of c-Mos protein for induction of activity (Sagata et al., 1989). The second peak appears in metaphase II and needs the resynthesis of cyclin (the second component of MPF complex) which is destroyed at the end of the first mitosis (Murray and Kirschner, 1989; Murray et al., 1989). Thus, MPF is a complex made of cyclin B and cdc2 proteins, which activation leads to a burst of protein phosphorylation 30–60 min prior to GVBD. This pleiotropic effect leads to activation of important molecules characterized as esential steps in the signal transduction pathways of several inducers of oocyte maturation.

After progesterone treatment, pre-existing maturation promoter factor (MPF, a complex of cyclin B and cdc2 protein) is activated prior to GVBD. This activation is sufficient to induce GVBD as evidenced by the fact that injection of active MPF induces precocious GVBD (Gotoh et al., 1991). Two peaks of MPF activity follows progesterone treatment of oocytes. The first coincides approximately with metaphase I. The second peak appears in metaphase II and remains stable until the oocyte is fertilized. MPF is activated by tyrosine dephosphorylation and threonine phosphorylation of cdc2. MPF, in turn, initiates a burst of protein phosphorylation 30–60 min prior to GVBD. This pleiotropic effect leads to execution of essential steps that lie in the signal transduction pathways of several inducers of oocyte maturation.

Progesterone-induced maturation requires the synthesis of Mos protein. Antisense ablation of c-*mos* mRNA blocks oocyte maturation and MPF activity induced by progesterone (Sagata et al., 1988). On the other hand, microinjection of Mos protein or mRNA induce MPF activity, GVBD and progression through meiosis I and II (Sagata et al., 1989; Yew et al., 1992). The protein synthesis requirement for MPF activation can be ascribed mainly to Mos translation, though recently it has been shown that the *de novo* synthesis of a cdc2 binding protein (cyclin B) also plays an important role (Nebreda et al., 1995).

A decrease in PKA activity has been also reported as a requirement for MPF activation and oocyte maturation. Progesterone activates cyclic AMP phosphodiesterase which in turn reduces the level of cAMP (Maller and Krebs, 1977), the activator of PKA. Inhibition of the decrease in cAMP levels (e.g. using IBMX) blocks progesterone-induced GVBD. Matten et al. (1994) have shown that PKA acts at multiple points to inhibit oocyte maturation through control of Mos synthesis, Mos-induced MPF activation and MPF-mediated activation of cdc25.

Increased synthesis of Mos kinase leads to MAP kinase activation via MEK. MEK is a direct target of Mos, and thus Mos acts as a MAPK kinase kinase (Nebreda and Hunt, 1993; Shibuya and Ruderman, 1993). Activation of the MAP kinase cascade by injection of a constitutively active MEK mutant may be sufficient for induction of MPF and GVBD (Kosako et al., 1994; Gotoh et al., 1995;

Haccard et al., 1995). Furthermore, using the MAPK phosphatase CL100, or antibodies that neutralize MAPK activity, it has been shown that MAPK activation is essential for progesterone- and Mos-induced maturation and MPF activation (Gotoh et al., 1995). Finally, injection of purified MPF into immature oocytes or the addition to interphase egg extracts activates endogenous MEK and MAPK (Ferrell et al., 1991; Gotoh et al., 1991). Based on these and other indications, it has been proposed that Mos, MAPK and MPF all fall in a feed-back loop that promotes maturation of *Xenopus* oocytes.

Moreover, it has been recently reported the Raf kinase is also involved in *Xenopus* oocyte maturation. Muslin et al. (1993) reported that the activation of Raf and p42^{MAPK} by progesterone was blocked by microinjection of *mos* antisense oligonucleotides, while microinjected *mos* mRNA activates the two kinases (Muslin et al., 1993). When *mos* antisense oligonucleotides were co-injected with v-Raf into oocytes, GVBD was induced and the p42^{MAPK} activated. The activation of this kinase seems to be necessary for progesterone- and Ras-induced GVBD. Finally, microinjection of a constitutively activated mutant of Raf-1 is able to induce itself maturation in oocytes (Muslin et al., 1993; Fabian et al., 1993). Thus, the authors concluded that Raf is downstream c-Mos activity. Activation of p42^{MAPK} by progesterone requires Mos expression (Fabian et al., 1993). Thus, c-Mos could be the link between MAP kinase activation and the MPF activation-pathways during cell division. Nishida and coworkers have shown that microinjection of purified MPF into immature oocytes leads to a dramatic activation of MEK1 and MAP kinase (Kosako et al., 1992). MPF could also induce activation of *Xenopus* MEK1 and MAP kinase *in vitro*. They conclude that MAP kinase lies downtream MPF. However, MEK1 treated with phosphatase 2A could not be reactivated by MPF *in vitro*, suggesting the existence of intermediary steps (Kosako et al., 1992). Thus, the Mos pathway can regulate the timing of MEK1 pathway activation, since the v-Raf in immature oocytes is sufficient to stimulate p42^{MAPK} (Muslin et al., 1993). Another argument supports this hypothesis: Raf activation in oocytes is very delayed compared to mammalian cells (Muslin et al., 1993). Oncogenic Ras activates p42^{MAPK} before MPF (Nebreda et al., 1993a) by a mechanism independent of c-Mos protein (Daar et al., 1991). Since direct interaction between Ras and Raf has been demonstrated (see Lacal and Carnero for a review), and *Xenopus* Raf protein seems to be necessary for Ras action, c-Mos may be a modulator of the MAP kinase phatway in *Xenopus* oocytes. A detailed review on Mos function in oocytes is discussed in Chapter 14.

Mitogenic signaling pathways lead to the phosphorylation of ribosomal protein S6 on Ser residues. In *Xenopus* oocytes this effect is made by the S6 kinase II, the homolog to mammalian rsk (Erikson, 1991), which is activated by phosphorylation by MAP kinase (Sturgill, 1988), also called ERK or p42^{MAPK}. The activation of p42^{MAPK} seems to be necessary for progesterone- and Ras-induced GVBD (Carnero et al., 1994). The p42^{MAPK} itself is also activated by direct phosphorylation on Thr and Ser by MAP kinase kinase (MEK) which is a substrate of Raf kinase (for a review see Lacal and Carnero, 1994). The activation of this kinase is necessary for progesterone- and Ras-induced GVBD. Finally, microinjection of a constitutively activated mutant of Raf-1 is able to itself induce maturation in oocytes (Muslin et al., 1993; Fabian et al., 1993). In this chapter, we describe some available techniques to analyze the activation of MPF, Raf-1, MAPK and S6 kinase in *Xenopus laevis* oocytes.

2   Materials

The materials required and procedures to obtain the oocytes have been described in detail in Chapter 11. Here only those novel experimental details specific for the analysis of intracellular kinases are described.

Solutions

- **BLO buffer (Solution 1)**, analysis of *in vitro* MPF activity is performed in the following buffer:
    20 mM Hepes pH 7.0
    10 mM β-glycerophosphate
    5 mM EGTA
    5 mM $MgCl_2$
    50 mM NaF
    2 mM DTT
    25 mg/ml aprotinin
    10 mg/ml leupeptine
    100 mM PMSF
- **Lysis buffer 1 (Solution 2)**, MAPK and Raf-l kinase assays are performed in the following buffer:
    50 mM Tris-HCl pH 7.5
    5 mM EDTA
    0.5% Triton X-100
    0.5% sodium deoxycholate
    10 mM $Na_4P_2O_7$
    50 mM NaF
    0.1 mM Na_3VO_4
    20 mg/ml leupeptin
    20 mg/ml aprotinin
    1 mM PMSF
- **T-TBS (Solution 3)**, Western blots of MAPK and Raf-l kinase are carried out in T-TBS:
    20 mM Tris-HCl pH 7.5
    150 mM NaCl
    0.005% Tween 20

- **Reaction buffer 1 (Solution 4)**
    20 mM Hepes pH 7.0
    5 mM β-mercaptoethanol
    10 mM MgC12
    10 μM γ-^{32}P-ATP (2–5 dpm/fmol)
    0.2 μg of PKA inhibitor
- **Lysis buffer 2 (Solution 5)**
    50 mM Tris-HCl pH 8.0
    2 mM EDTA
    10% glycerol
- **Hepes buffer (Solution 6)**
    20 mM Hepes pH 7.0
    5 mM β-mercaptoethanol
    10 mM MgC12
- **Blocking buffer (Solution 7)**
    25 mM Tris-HCl pH 7.5
    0.05% Tween
    150 mM NaCl
    5% BSA
- **Lysis buffer 3 (Solution 8)**
    50 mM β-glycerophosphate pH 7.3
    1.5 mM EGTA
    1 mM DTT
    400 μM PMSF
    2 μM leupeptin
    25 μg/ml aprotinin
    5 mM NaPPi
    1 mM NaF
- **Reaction buffer 2 (Solution 9)**
    50 mM Tris pH 7.4
    1 mM DTT
    10 mM MgC1$_2$
    50 μM [γ-^{32}P]-ATP (3000 Ci/mmol)
    2.5 μM PKA inhibitor
- **Reaction buffer 3 (Solution 10)**
    250 μM rsk-substrate peptide (AKRRRLSSLRA, Upstate Biotechnology
    Inc., #17–136)
    50 mM glycerophosphate pH 7.3
    7 mM NaF
    0.3 mM EDTA
    150 nM MgC12
    2 mM DTT
    50 μM [γ-^{32}P]-ATP (3000 Ci/mmol, Amersham)
    7 μM PKA inhibitor peptide (Sigma)

3   Methods

For preparation of the oocytes and micropipets, as well as the injection procedure and the oocyte maturation assay (GVBD assay), see Chapter 11.

3.1 Analysis of MPF activity

Analysis of oocyte maturation can be determined by means of biochemical markers such as the activation of specific intracellular enzymes. The cdc2 kinase is broadly used for this purpose since it is one of the components of the **Maturation Promoting Factor (MPF)**. Several alternative protocols to evaluate the activation of cdc2 kinase are available. Here, we will describe two of them: l) determination of the *in vitro* MPF activity in whole oocyte extracts; 2) determination of the *in vitro* MPF activity after specific precipitation with pl3^{suc} bound to Sepharose beads. Series of at least 20 oocytes are treated or microinjected with the compounds under investigation. At the desired times, whole oocytes are transferred to a microcentrifuge tube and buffer is completely removed. Oocytes are then homogenized using the tip of a 1 ml Eppendorf disposable pipet and after homogenization, resuspended in a small volume of BLO buffer.

Whole extract protocol

Protocol 1   Determination of the *in vitro* MPF activity in whole oocyte extracts

1. Samples prepared as above (3.1) are centrifuged at 13000 g for 15 min to eliminate insoluble material and supernatants are assayed for 10 min at 30°C in a final reaction volume of 50 µl of Reaction buffer 1 (**Solution 4**). As substrate, 1 mg/ml of type I histone (Boehringer Mannheim) is added.

2. Stop reaction by addition of PAGE-sample buffer and keep samples on ice.

3. Boil samples for 2 min and resolve them by a 15% polyacrylamide gel electrophoresis (PAGE).

4. Dry the gel and expose several hours at -70°C to a sensitive autorradiographic film.

To estimate the radioactivity incorporated into the substrate (histone) each band can be excised from the gel and counted in scintillation liquid. Other equivalent methods for quantitation of radioactivity in the gel can also be used such as an electronic autoradiography system. A typical experiment where activation of cdc2 is followed by this protocol is shown in Figure 1A.

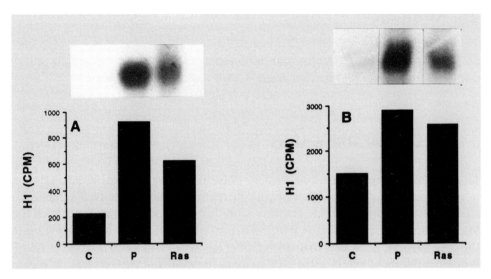

Figure 1   Activation of MPF by progesterone and *ras*-p21.

Phosphorylation of H1 by MPF, performed as described in the text, was carried out after progesterone treatment or microinjection of the activated Ras protein, using either total oocyte extracts (**A**) or p13-agarose precipitates (**B**). Unstimulated, control oocytes (**C**), oocytes treated with 1 μg/ml progesterone (**P**) or oocytes microinjected with 25 ng of Ras p21 (Ras).

Precipitation protocol

The cdc2 protein kinase can be precipitated by using pl3^{suc}, a protein that specifically binds to it. pl3^{suc} can be purified from a recombinant source (Brizuela et al., 1987).

Protocol 2   Determination of the *in vitro* MPF activity after specific precipitation with pl3^{suc} bound to Sepharose beads

1. The *E.coli* strain BL21.DE3 containing the pRK172.sucl+ plasmid is grown in 400 ml of LB-broth containing 100 µg/ml ampicillin, and pl3^{suc} expression induced by 0.4 mM IPTG.

2. Bacterial cultures are resuspended in lysis buffer 2 (**Solution 5**) and the soluble fraction is loaded on a 1.5 × 80 cm Sepharose CL6B column.

3. The sample is eluted in a gradient from 0 to 500 mM NaCl made in the lysis buffer 2 (**Solution 5**) without glycerol.

4. Fractions of 2 ml are collected and a small aliquot of each analyzed by SDS-PAGE.

5. pl3^{suc} is linked to sheparose CNBr-4B according to the protocol provided by the manufacturer (Pharmacia).

6. MPF assay is performed after precipitation using pl3^{suc}-sepharose beads.

7. After treatment with the compounds under investigation or microinjections, oocytes are lysed, resuspended in BLO buffer (**Solution 1**), and centrifuged at 13000 g as described above.

8. The resulting supernatants are incubated for 2 h under constant agitation at 4°C with 50 µl of the pl3^{suc}-sepharose solution in a final volume of 1 ml.

9. The pl3^{suc}-Sephasore pellets are washed once with BLO (**Solution 1**) and twice with Hepes buffer (**Solution 6**).

The kinase assays using the pl3^{suc}-Sepharose precipitates are performed and quantified as described above for total extracts. This protocol has the advantage that it provides a more specific assay due to the specific binding of pl3^{suc} to the cdc2 kinase. A typical experiment where activation of cdc2 is followed by this protocol is shown in Figure 1B.

3.2   Assay for activation of Raf-1 kinase

The Raf-1 kinase can be assayed by alternative protocols such as the ability to phosphorylate its substrate, the MEK, or by its mobility shift on a PAGE using specific antibodies to the Raf-1 kinase. We describe here only the second procedure. For this protocol, we use a polyclonal antibody raised in our laboratory. Generation of the Raf-1 polyclonal antibody was achieved by standard procedures using the peptide CTLTTSPRLLPVF, which provides specificity for Raf-1, conjugated to tyroglobulin by cross-linking with glutaraldehyde. The conjugated peptide was used to immunize rabbits and the generated antiserum was analyzed by Western blot. The antiserum was found to recognize in whole cell lysates only one band corresponding to the Raf-1 kinase.

Protocol 3   Assay for activation of Raf-l kinase

1. After treatment of the oocytes with the desired substance or microinjections, terminate incubations by homogenization of the oocytes as described above for the MPF assay (Section 3.1).

2. Add 300 µl of ice-cold lysis buffer to each sample.

3. Remove nuclei and detergent-insoluble material by centrifugation at 10000 g for 10 min.

4. The resulting supernatants are assayed for estimation of total cell protein (Bio-Rad) and equal amounts of cell lysate (typically 40 µg) are boiled at 95°C for 5 min in SDS-PAGE sample buffer.

5. Proteins are resolved onto 8% SDS-PAGE gels poured in 20×20 cm glasses.

6. Resolved proteins are transferred to nitrocellulose paper.

7. The resulting blots are blocked for 2 h in 2% non-fat dried milk in T-TBS, then washed once in T-TBS and incubated 4 h with a 1:1000 dilution of the polyclonal anti-Raf-l kinase antibody.

8. Wash blots three times for 10 min in T-TBS (**Solution 3**), incubate 1 h with 1:1000 anti-rabbit Ig biotinylated (Amersham), and wash three times for 10 min with T-TBS.

9. Blots are then incubated 30 min with streptavidin horseradish peroxidase (Amersham) 1:1000 in T-TBS (**Solution 3**). After washing three times with T-TBS 10 min. Raf-l kinase is detected by the ECL system (Amersham).

Activation of the Raf-l kinase is determined by the mobility shift (retardation) produced as a consequence of its phosphorylation. A typical example is shown in Fig. 2.

3.3   Analysis of MAPK activation

There are several alternative protocols for the determination of the activation of MAPK. Most of these assays rely upon specific recognition of the MAPK by antibodies raised either against this kinase or capable of recognizing P-Tyrosine residues. Alternatively, MAPK activity can be determined by an *in situ* kinase assay or after partial purification.

   For Western blot determinations we have used either a polyclonal antibody raised in the laboratory against MAPK or a commercial anti-P-Tyr antibody, but many companies sell excellent antibodies against both MAPK or P-Tyr. Generation of the MAPK polyclonal antibody was achieved by standard procedures using a peptide corresponding to the C-terminus of MAPK (KERLKELIF-QETAR) conjugated to tyroglobulin by cross-linking with glutaraldehyde. The

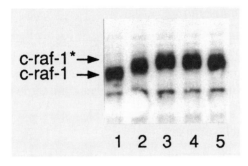

c-raf-1*→
c-raf-1 →

1 2 3 4 5

Figure 2 Activation of Raf-1 by microinjection of Ras proteins.

Activation of the Raf1 kinase was performed as described in the text using a specific antibody against Raf in control oocytes (1) and in oocytes injected with 10 ng/oocyte (2), 20 ng/oocyte (3), 50 ng/oocyte (4), 100 ng/oocyte (5) of purified Ras protein. Activation of the Raf1 kinase can be observed by the mobility shift (retardation) produced as a consequence of its phosphorylation (c-raf-1*).

conjugated peptide was used to immunize rabbits and the generated antiserum was analyzed by Western blot. The antiserum was found to recognize in whole mammalian cell lysates only two bands (p44 and p42) corresponding to both isoenzymes of the MAPK family, and just the p42XeMAPK in *Xenopus laevis* oocytes extracts. The specific protocols for the alternative assays for MAPK are described below.

Assay for MAPK based on mobility shifts

Protocol 4   Assay for MAPK based on mobility shifts

1. Oocytes are treated with the substance under investigation or microinjected. After the required period of time, oocytes are homogenized as described above for the MPF assay (Section 3.1).

2. Add 300 µl of ice-cold lysis buffer to each sample.

3. Nuclei and detergent-insoluble material are removed by centrifugation at 10000 g for 10 min. The resulting supernatants are assayed for estimation of total cell protein (Bio-Rad) and equal amounts of cell lysate (typically 40 µg) are boiled at 95°C for 5 min in SDS-PAGE sample buffer.

4. For Western blot analysis, proteins are electrophoresed onto 10% SDS-PAGE gels poured in 20×20 cm glasses. Separated proteins are transferred to nitrocellulose and blots are blocked for 2 h in 2% non-fat dried milk in T-TBS. Blots are washed once in T-TBS (**Solution 3**) and incubated 4 h with a 1:1000 dilution of the polyclonal anti-MAPK antibody. Blots are washed three times for 10 min in T-TBS, incubated 1 h with 1:1000 anti-rabbit Ig biotinylated (Amersham), washed three times for 10 min with T-TBS and incubated 30 min with streptavidin-horseradish peroxidase (Amersham) 1:1000 in T-TBS. After washing three times with T-TBS 10 min both MAPK enzymes are detected by the ECL system (Amersham).

The activation of MAP kinases in response to different stimuli is assessed by the mobility shift produced as a consequence of the hyperphosphorylation of these kinases. A typical example for this procedure is shown in Figure 3A.

MAPK assay determined by tyrosine phosphorylation

Protocol 5   MAPK assay determined by tyrosine phosphorylation

1. Oocytes are treated with the substance under investigation or microinjected. After the required period of time, oocytes are homogenized as described above for the MPF assay (Section 3.1).

2. Add 300 μl of ice-cold lysis buffer to each sample.

3. Nuclei and detergent-insoluble material are removed by centrifugation at 10000 g for 10 min. The resulting supernatants are assayed for estimation of total cell protein (Bio-Rad) and equal amounts of cell lysate (typically 40 μg) are boiled at 95°C for 5 min in SDS-PAGE sample buffer.

4. Proteins are resolved onto 10% SDS-PAGE gels poured in 20×20 cm glasses. Resolved proteins are transferred to nitrocellulose and filters are blocked in buffer (**Solution 7**) for 2 h at 50°C. Phosphotyrosine containing proteins are detected by incubating the blot for 2 h in the same buffer with 1:500 dilution of antiphosphotyrosine specific antibody (Upstate Biotechnology, Inc.). Blots are washed three times for 10 min in T-TBS (**Solution 3**), incubated 1 h with 1:1000 anti-mouse Ig biotinylated (Amersham), washed three times for 10 min with T-TBS and incubated 30 min with streptavidin-horseradish peroxidase (Amersham) 1:1000 in T-TBS. After washing three times with T-TBS 10 min MAPK activation is detected by the ECL system (Amersham).

The activation of MAP kinases in response to different stimuli is assessed by the appearance of specific bands showing the P-Tyr forms of these kinases, as shown in Figure 3A.

MAPK assay by chromatography on a Mono-Q column
Finally, MAPK can be determined by a more elaborate protocol after a preliminary purification of the MAPK enzymes through a Mono-Q column. We have succesfully used the following procedure from oocytes extracts.

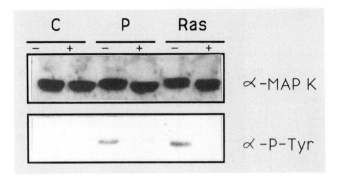

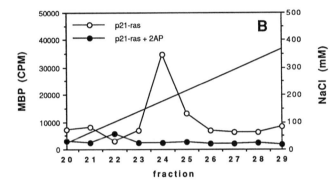

Figure 3 Activation of MAP Kinases by progesterone treatment or microinjection of the Ras protein.

(A) Mobility shift and tyrosine phosphrylation: Gel electrophopresis mobility (upper gel) and tyrosine phosphorylation (lower gel) state of p42^{MAPK} after progesterone or *ras*-stimulation. Western blots from total extracts of oocytes treated (+) or not (-) with 10 mM 2-Amino purine (2-AP) were performed as described in the text. C: untreated; P: Progesterone 1 μg/ml; Ras: 25 ng/oocyte. (B) Analysis of MAP kinase activity after partial purification of the enzyme by chromatography on a Mono-Q column. Oocytes treated (●) or not (○) with 10 mM 2-AP were microinjected with 25 ng of v-H-*ras*-p21. 20 h after microinjection, oocytes were homogenized and processed as described in the text, and fractions assayed for MAP kinase activity with MBP as substrate.

Protocol 6   MAPK assay by chromatography on a Mono-Q column

1. After treatment or microinjection, oocytes are lysed in **Solution 8**.

2. Extracts are cleared at l00000 g in a TL100 centrifuge and filtered in 0.2 μm filters.

3. Samples of 2 mg of total protein are applied to a Mono Q column that has been equilibrated in the same buffer (**Solution 8**) without NaF.

4. The columns are washed in the equilibrating buffer (**Solution 8**).

5. Proteins are eluted with a linear gradient of NaCl (0 to 500 mM). Fractions of 2 ml are collected.

6. Aliquots of 30 μl from each fraction are assayed for MAPK activity using 0.25 mg/ml MBP as substrate in the Reaction buffer 2 (**Solution 9**) in a final volume of 50 μl.

7. After 15 min at 30°C samples are spotted onto Watman p81 phosphocellulose paper filters, washed extensively with 1% ortophosphoric acid and once with 95% ethanol.

The radioactivity retained on the filters is then quantified in a scintillation counter. A typical experiment is shown in Figure 3B.

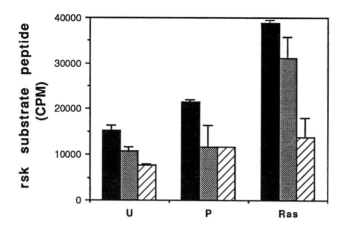

Figure 4 Effect of inhibition of MPF or p42^{MAPK} on S6 kinase activity.

Oocytes were treated with 1 μg/ml of progesterone (P), microinjected with 25 ng of Ras-p21 (Ras) or left untreated (U) in Ringer's buffer alone (■), in Ringer's buffer containing 2 μM CHX (▨) or 10 mM 2-AP (⧄).

3.4   S6 Kinase assay

Activation of the S6 kinase II enzyme is determined by analysis of the activity in whole extracts using as a substrate a specific peptide.

Protocol 7   S6 Kinase assay

1. Series of 20 oocytes treated or microinjected with the substance under investigation are homogenized as described for the MAPK assay (Section 3.1) and resuspended in a final volume of 20 μl of BLO buffer (**Solution 1**).

2. Extracts are cleared at 13000 rpm for 15 min to eliminate insoluble materials.

3. Aliquots of 30 μl of the resulting supernatants are assayed with the commercial peptide AKRRRLSSLRA (Upstate Biotechnology Inc., #17–136) which is specific for the S6 KII (8). Mix aliquots with Reaction buffer 3 (**Solution 10**) in a final volume of 50 μl.

4. The assays are incubated at 30°C for 20 min and stopped with ice-cold TCA to a final concentration of 16% TCA.

5. Samples are kept 15 min at 4°C and centrifuged at 13000 rpm for 15 min.

6. The supernatants are spotted onto Whatman p81 phosphocellulose paper filters, washed extensively with 1% ortophosphoric acid and once with 95% ethanol.

7. The radioactivity retained on the filters are quantified in a scintillation counter.

A typical autorradiography for RAF kinase activation followed by the mobility shift assay is shown in Figure 4.

Acknowledgements

This publication has been possible thanks to the specific support from DGICYT (Grant #PB94–0009) and FIS (Grant #96/2136). AC is the recipient of an EMBO Long-Term Fellowship.

References

Brizuela L, Draetta G, Beach D (1987) pl3sucl acts in the fission yeast cell division cycle as a component of thep34cdc2 protein kinase. *EMBO* J. **6**: 3507–3514.

Carnero A, Jimenez B, Lacal JC (1994) Progesterone but not Ras requires MPF for *in vivo* activation of MAP K and S6 KII. MAP K is an essential conexion point of both dignaling pathways. *J. Cell. Biochem*. 55: 465–476.

Daar I, Nebreda AR, Yew N, Sass P, Paules R, Santos E, Wigler M, Vande Woude GF (1991) The ras oncoprotein and M-phase activity. *Science* 253: 74–76.

Dunphy GW, Brizuela L, Beach D, Newport J (1988) The *Xenopus* cdc2 protein is a component of MPF, a cytoplasmic regulator of mitosis. *Cell* 54: 423–431.

Erikson RL (1991) Structure, expression, and regulation of protein kinases involved in the phosphorylation of ribosomal protein S6. *J. Biol. Chem*. 266: 6007–6010.

Fabian JR, Morrison DK, Daar IO (1993) Requirement for Raf and MAP kinase function during the meiotic maturation of *Xenopus* oocytes. *J. Cell Biol*. 122: 645–652.

Ferrell JE, Wu M, Gerhart JC, Martin GS (1991) Cell cycle tyrosine phosphorylation of p34cdc2 and microtubule associated protein kinase homolog in *Xenopus* and eggs. *Mol. Cell. Biol*. 11: 1965–1971.

Gotoh Y, Masuyama N, Dell K, Shirakabe K, Nishida E (1995) Initiation of *Xenopus* oocyte maturation by activation of the mitogen activated protein kinase cascade. *J. Biol Chem*. 270: 25898–25904.

Gotoh Y, Moriyama K, Matsuda S, Okumura E, Kishimoto T, Kawasaki H, Suzuki K, Yahara I, Sakai H, Nishida E (1991) *Xenopus* M-phase MAP kinase: isolation of its cDNA and activation by MPF. *EMBO*. J. 10: 2661–2668.

Haccard OA, Lewellyn A, Hartley RS, Erikson E, Maller JL (1995) Induction of *Xenopus* oocyte maturation by MAP kinase. *Dev. Biol*. 168: 677–682.

Kameshita I, Fujisawa H (1989) A sensitive method for detection of calmodulin-dependent protein kinase II activity in sodium dodecyl sulfate polyacrylamide gel. *Anal. Biochem*. **183**: 139–143.

Kosako H, Gotoh Y, Nishida E (1994) Requirement for the MAPKinase kinase/ MAPkinase cascade in *Xenopus* oocyte maturation. *EMBO J* 13: 2131–2138.

Kosako H, Gotoh Y, Matsuda S, Ishikawa M, Nishida E (1992) *Xenopus* MAP kinase activator is a serine/threonine/tyrosine kinase activated by threonine phosphorylation. EMBO J. 11: 2903–2908.

Maller JL, Krebs EG (1977) Progesterone-stimulated meiotic cell division in *Xenopus* oocytes. Induction by regulatory subunit and inhibition by catalytic subunit of adenosine 3':5'-monophosphate-dependent protein kinase. *J Biol Chem* 252: 1712–1718.

Matten WI, Daar I, Vande Woude GF (1994) Protein kinase A acts at multiple points to inhibit *Xenopus* oocyte maturation. *Mol Cell Biol* 14: 4419–4426.

Murray AW, Solomon MJ, Kirschner MW (1989) The role of cyclin synthesis and degradation in the control of maturation promoting factor activity. *Nature* 339: 280–286.

Murray AW, Kirschner MW (1989) Cyclin synthesis drives the early embryonic cell cycle. *Nature* 339: 275–280.

Muslin AJ, MacNicol AM, Williams LT (1993) Raf-1 protein kinase is important for progesterone-induced *Xenopus* oocyte maturation and acts downstream of mos. *Mol Cell Biol* 13: 4197–4202.

Nebreda AR, Hunt T (1993) The c-mos proto-oncogene protein kinase turns on and maintains the activity of MAP kinase, but not MPF, in cell-free extracts of *Xenopus* oocytes and eggs. *EMBO J* 12: 1979–1986.

Nebreda AR, Gannon JV, Hunt T (1995) Newly synthesized protein(s) must associate with p34cdc2 to activate MAP kinase and MPF during progesterone-induced maturation of *Xenopus* oocytes. *EMBO J* 14: 5597–5607.

Nebreda AR, Porras A, Santos E (1993) p21-ras induced meiotic maturation of *Xenopus* oocytes in the absence of protein synthesis: MPF activation is preceded by activation of MAP ans S6 kinases. *Oncogene* 8: 467–477.

Posada J, Yew N, Ahn NG et al. (1993) Mos stimulates MAP kinase in *Xenopus* oocytes and activates a MAP kinase kinase *in vitro*. Mol. Cell. Biol. *13:* 2546–2553.

Sagata N, Daar I, Oskarsson M, Showalter SD, Vande Woude GF (1989) The product of the mos proto-oncogene as a candidate "initiator" for oocyte maturation. *Science* 245: 643–646.

Sagata N, Oskarsson M, Copeland T, Brumbaugh J, Vande Woude GF (1988) Function of c-mos proto-oncogene product in meiotic maturation in *Xenopus* oocytes. *Nature* 335: 519–525.

Shibuya EK, Ruderman JV (1993) Mos induces the *in vitro* activation of mitogen-activated protein kinases in lysates of frog oocytes and mammalian somatic cells. *Mol Biol Cell* 4: 781–790.

Sturgill TW, Ray LB, Erikson E, Maller JL (1988) Insulin-stimulated MAP-2 kinase phosphorylates and activates ribosomal protein S6 kinase II. *Nature* 334: 715–718.

Yew N, Mellini ML, Vande Woude GF (1992) Meiotic initiation by the *mos* protein in *Xenopus*. Nature *355*: 649–652.

13 Signal Transduction in *Xenopus laevis* Oocytes by *Ras* Oncoproteins and Lipid Metabolites

A. Carnero and J.C. Lacal

Contents

1 Introduction

The mammalian *ras* genes (H-*ras*, K-*ras* and N-*ras*) encode a family of closely related 21 kDa proteins that are members of the superfamily of GTP binding proteins (G-proteins). Ras proteins participate in a number of intracellular signalling pathways, and the conservation of Ras throughout the evolution of eukaryotes suggests that this participation is critical.

Like other GTPases, Ras proteins are regulated by nucleotide binding. Ras is active when bound to GTP and inactive when bound to GDP (Fig. 1). The conversion of inactive Ras into an active state requires the exchange of GDP for GTP and is stimulated by nucleotide exchange factors (GEFs). Inactivation of Ras via GTP hydrolysis is stimulated by GTPase activating proteins (GAP, neurofibromin). Antibodies to Ras or point mutations in Ras proteins that prevent exchange of GDP for GTP inhibit its function. In contrast, mutations that block the GTPase activity, or increase the exchange of GDP for GTP result in chronic activation. Constitutive Ras activation can contribute to cellular transformation, and mutations which accomplish this feat have been found in human tumors, particularly at amino acids 12,13 and 61 (Bos, 1989; Leon and Pellicer, 1992).

Signaling via Ras proteins also requires membrane localization. Ras proteins are targeted to the plasma membrane by a complex processing which involves farnesylation of a cysteine residue near the c-terminus (Hancock and Marshall, 1993). Alternative sequences such as polybasic residues at the C-ter-

Methods and Tools in Biosciences and Medicine
Microinjection, ed. by J. C. Lacal et al.

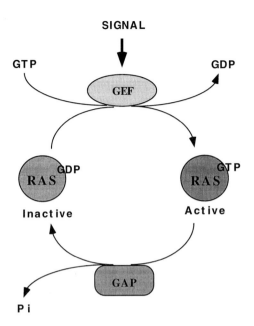

Figure 1   Biochemical cycle of Ras proteins. Activation of the Ras protein is achieved by the exchange of GDP for GTP. Inactivation of the protein is mechiated by hydrolysis of GDP.

minus (Hancock and Marshall, 1993) or the myristoilation signaling peptide from c-Src (Lacal et al., 1988), also renders active Ras proteins. On the other side, mutants unable to locate to the membrane are not biologically active.

The Ras proteins couple a variety of cell surface receptors to intracellular second messengers. Growth factor receptors possessing tyrosine kinase activity promote an increase in GTP-bound Ras proteins and induce Ras activation. The inhibition of Ras function through neutralizing antibodies or by expression of a dominant negative Ras mutant (i.e., Ras Asn[17]) can block the mitogenic activity of several growth factors (PDGF, EGF, NFG, insulin).

The way in which Ras passes signals to downstream effectors is incompletely understood. One common mechanism occurs through the activation of the serine/threonine kinase, Raf. GTP-bound Ras binds to inactive, cytoplasmic Raf and directs its translocation to the plasma membrane. Membrane associated Raf is subsequently activated by an unknown mechanism. This eventually leads to the activation of MAP kinase (MAPK) by MAPK kinase (MEK). The activation of MAPK leads to phosphorylation and activation of transcription factors that promote the transcription of genes required for entry into S phase.

Raf-independent Ras pathways have also been described (Fig. 2). Ras activation can induce membrane ruffling of fibroblasts and reorganization of the cytoskeleton by mechanisms involving Rac1 and the Rho family of GTPases, respectively. Moreover, a Raf-independent mitogenic pathway that is mediated by the Rho and Rac proteins has been described (Qiu et al., 1995; Denhardt, 1996).

Ras can also signal by activating PI3 kinase through direct protein-protein interactions (Rodriguez-Viciana et al., 1996). PI3 kinase promotes the genera-

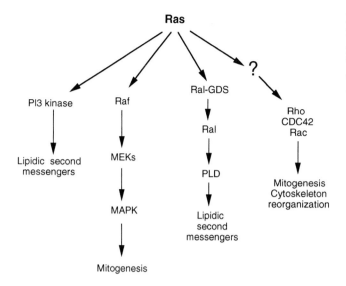

Figure 2 Some of the signal transduction pathways activated by Ras.

The tree best characterizal effectors for Ras proteins are depicted.

tion of lipid second messengers that have been found to activate some novel kinases (PKB/Akt). Finally, Ras can also activate phospholipase D (PLD) both in *Xenopus* oocytes (Carnero and Lacal, 1995) and mammalian cells (Carnero et al., 1994b). This effect is though to be mediated by binding of Ras to Ral GDS and the subsequent activation of the Ral pathway (Jiang et al 1995). This pathway constitutes a Ras-dependent signal which is distinct from and cooperates with the Raf-MAPK pathway (for review see Pazin and Williams, 1993; Khosravi-far and Der, 1994; Kantz and McCormick, 1997).

2 *Xenopus* oocytes as a tool to study the biological activity of Ras proteins

In 1985, Wigler's group showed that injection of purified, recombinant human H-Ras protein induced maturation of *Xenopus* oocytes (Birchmeier et al., 1985). The oocyte processed bacterially-produced mammalian Ras proteins in a manner similar, if not identical, to that in which mammalian cells process the endogenous protein (Zhao et al., 1994). Meiotic maturation can also be accomplished by injection of Ras mRNA (Johnson et al., 1990).

Analysis of the biological activity of different Ras mutants shows an exact correlation between their activity in mammalian cells and their activity in the *Xenopus* assay (see Table 1). The oncogenic Ras Val^{12} protein is 100-fold more potent than wild type Ras in inducing GVBD (Birchmeier et al., 1985). Other constitutively active mutants (e.g. Ras Lys^{12} or Ras Leu^{61}) also induced maturation. However, Ras proteins mutated in the effector loop (aminoacids 32–40) (e.g. Ras Glu^{38} or Ras Val^{12}Ala^{35}) did not. Ras proteins that are constitutively activated but fail to localize to the membrane (e.g. Ras Val^{12}Ser^{186}) are also unable to induce maturation (Gibbs et al., 1989; Lacal, 1990; Sadler et al., 1990; Pomerance et al., 1992).

Table 1   Effect of microinjection of different mutants of the Ras protein in Xenopus oocytes.

Ras protein	Features	GVBD	Mammalian cell transformation
Gly 12	wild type	+	+
Val 12	Low GTPase activity	+++	+++
Lys 12	Low GTPase activity	+++	+++
Leu 61	Low GTPase activity	+++	+++
Val 12–Thr 59	Low GTPase activity and Increased nucleotide exchange rate	+++	++++
Glu 38	No interaction with the effector	–	–
Val 12 Ala 35	No interaction with the effector	–	–
Leu 36 Thr 59	No interaction with the effector	–	–
Lys 12 Glu 38	No interaction with the effector	–	–
Val 12 Ser 186	No binding to the membrane	–	–
Val 12 Thr 59 Δ174 term	No binding to the membrane	–	–

The GVBD assay has been used as a tool in pharmacological assays of Ras inhibition, especially in the case of inhibitors such as peptides that fail to penetrate cell membranes. Gibbs et al. (1989) showed that a Ras mutant that is unable to bind to the membrane inhibits the Ras Val^{12} induced GVBD in trans. Similarly, peptides that inhibit farnesyl transferase block GVBD by preventing the Ras localization to the membrane (Zhao et al., 1994; Garcia et al., 1993). Similar effects have been noted with peptides from the GAP-binding domain of Ras (Chung et al., 1991; Losardo et al., 1995) and with the antibiotic azatyrosine (Campa et al., 1992). Other proteins such as Rap 1A and Rap 1B that have been found to block the interaction of Ras with its GTPase activating protein *in vitro*, also inhibit the Ras-induced maturation (Campa et al., 1991) as does a Ras-interacting domain of the RalGDS family members (Koyama et al., 1996). Some Ras inhibitors, as well as Ras antibodies that block the effector domain, inhibit insulin- or IGF-1-induced maturation of *Xenopus laevis* oocytes (Korn et al., 1987; Desphande and Kung, 1987; Soto-Cruz and Nagee, 1995). However, none of these Ras-inhibitors affect the maturation induced by progesterone. This indicates that Ras proteins are an essential step in the insulin- or IGF-1-signal transduction pathway leading to GVBD, but that progesterone induces maturation independently of Ras signaling (Fig. 3).

3   Signal transduction pathways induced by Ras activation in *Xenopus* oocytes

Ras oncoproteins induce a maturation pathway that differs from the progesterone-induced pathway, although some biochemical responses are shared. As do progesterone, insulin and IGF-1, oncogenic Ras proteins stimulate a cAMP phosphodiesterase that down-regulates PKA activity by reducing cAMP levels. Inhibitors of the phosphodiesterase (e.g. IBMX) also inhibit Ras-induced GVBD (Sadler and Maller, 1989). However Ras-induced maturation differs from progesterone induced maturation in that the inhibitory effects of PKA on the Ras pathway are not related to inhibition of Mos protein since Ras-induced maturation does not require this protein (Daar et al., 1991). Thus, the Ras pathway also differs in the lack of a requirement for *de novo* protein synthesis (Allende et al., 1988, Nebreda et al., 1992; Carnero et al., 1994c).

Both Ras oncoproteins and progesterone treatment induce the activation of MAP kinase. However, with Ras, this activation persists even in the presence of protein synthesis inhibitors (Nebreda et al., 1992; Carnero et al., 1994c) and in oocyte extracts (Fukuda et al., 1994). This suggest that activation of MAP kinase by Ras and progesterone occurs through two different pathways. GTP-Ras binds to Raf in the cytosol and targets Raf to the membrane where it becomes activated. The importance of this step is emphasized by the fact that the addition of the Ras membrane localization signal to Raf is sufficient to direct Raf to the membrane and to induce its activation (Leevers et al., 1994; Stokoe et al., 1994). Moreover, Ras mutants in Ser 186 inhibit Ras signaling by trapping Ras and Raf in a stable complex in the cytosol (Miyake et al., 1996). Raf, as well as another *Xenopus* protein, REKS (Shibuya et al., 1992), phosphorylate MEK (Macdonald et al., 1993). Thus, activation of Raf by Ras leads to the activation of the MAPK pathway through MEK (Fig. 4). Microinjection of constituitively active Raf or MEK proteins induce meiotic maturation of *Xenopus* oocytes (Kosako et al., 1994; Gotoh et al., 1995; Haccard et al., 1995). As with progesterone treatment, microinjection of dominant-negative mutant of Raf, MEK or MAPK, or the MAPK phosphatase CL100, inhibits the Ras-dependent maturation (Fabian et al., 1993; Alessi et al., 1993).

The activation of MAPK induces the activation of other proteins such as S6KII (the *Xenopus* homolog of mammalian p90rsk) or, indirectly, MPF. In progesterone-treated oocytes, MAPK activation is parallel to MPF activation. However, in the Ras-induced pathway, the MAP kinase activation precedes by several hours MPF activity (Nebreda et al., 1992; Carnero et al., 1994c), probably indicating the absence of a positive-feedback loop such as has been proposed for progesterone pathway (Fig. 3).

Activation of the Raf-MAPK pathway is a key event in Ras induced GVBD. Agents that raise cAMP levels inhibit Ras-dependent mitogenesis, mammalian cell transformation and GVBD (Sadler and Maller, 1989; Sadler et al., 1990; Cook and McCormick, 1994; Wu et al., 1994; Graves et al., 1994). It has been

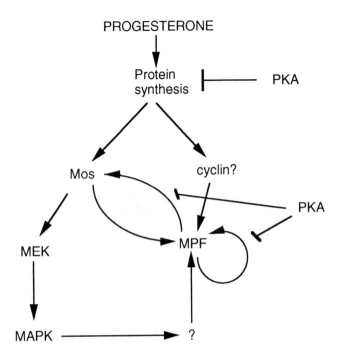

Figure 3 Kinase cascades activated by progesterone in Xenopus laevis oocytes leading to maturation.

Interrelationship among different cascades activated after progesterone treatment is represented. The involvement of PKA activation blocking these pathways is also shown.

shown that activation of PKA blocks the Raf-MAP kinase pathway by inhibiting the activation of Raf in two different ways. Phosphorylation of Raf by PKA strongly decreases the ability of Raf to associate with GTP-Ras (Kikuchi and Williams, 1996) and also interferes with the phosphorylation and activation of Raf by protein kinases as PKC or Lck (Hafner et al., 1994).

The signaling pathway leading from Ras to GVBD is very similar to the pathway leading from Ras to DNA synthesis. However, there is at least one important difference. In mammalian fibroblasts, depletion of PKC by chronic treatment by phorbol esters or the inhibition of PKC by specific inhibitors as bisindolylmaleimide (a staurosporine analog specific for the classical and the novel PKC isoforms) inhibits the Ras- but not serum-induced DNA synthesis (Lacal et al., 1987b). Microinjection of purified PKC restores the response (Lacal et al., 1987b). However, in the oocyte, inhibition of PKC does not alter maturation induced by oncogenic Ras (Carnero and Lacal, 1993; Carnero et al., 1994b), suggesting that only some Ras effects depend on PKC.

4   Ras-induced phospholipid alterations in the oocyte

The microinjection of activated Ras proteins into oocytes induces a rapid generation of DAG from hydrolysis of phosphatidylcholine (PC) (Lacal et al., 1987a; Lacal, 1990). Generation of DAG correlates with the ability of the Ras proteins to induce GVBD. These results agree with previous observation in NIH3T3 cells transformed with *ras* oncogenes where transforming activity correlated with elevated levels of DAG, but not with phosphatidylinositol (PI) hydrolysis (Lacal et al., 1987c). DAG could be produced from PC by hydrolysis mediated by a phospholipase C (PLC), generating DAG and phosphorylcholine or by phospholipase D (PLD) generating phosphatidic acid (PA) and choline which are further converted to DAG and phosphorylcholine by a PA hydrolase and a choline kinase, respectively (Fig. 5). However, both DAG and phosphorylcholine can be used in other metabolic pathways, such as the sphingomyeline cycle, making it difficult to pinpoint the origin of these compounds.

Microinjection of activated Ras proteins in *Xenopus* oocytes induces a rapid rise in the level of PA which is not sensitive to the DAG kinase inhibitor R59023 (Carnero and Lacal, 1995). If DAG release is a consequence of the consecutive activation of PLD and PA hydrolase, the inhibition of the latter enzyme should drastically affect the DAG levels. Using propanolol, an inhibitor of PA hydrolase that does not affect the activities of PLD or PLC (Koul and Hauser, 1987), we found that DAG remains at basal levels after Ras injection, indicating that the DAG is produced from the PA hydrolysis (Carnero and Lacal, 1995). Moreover, direct activation of PLD can be probed through its ability to use alcohol (usually ethanol or butanol) instead of H_2O in the hydrolysis of PC to produce a non-metabolizable compound. This is a rather specific activity for PLD designated as its transphosphatidylation activity. Activated Ras proteins induce a rapid increase in the activity of PLD but not DAG kinase (Carnero and Lacal, 1995). These results are in agreement with what has been found in mammalian cells where *ras* oncogenes also activate a PLD rather than a PLC (Carnero et al., 1994a). Activation of PLD by *ras* occurs through the activation of Ral (Jiang et al 1995) and is independent of PKC (del Peso et al., 1996; del Peso et al., 1997). Finally, it has been reported that Ras proteins mediate the activation of PLD by c-*Src* (Song et al., 1991) and that more than a single mechanism for PLD activation coexist in mammalian cells, one is dependent on Ras and another dependent on PKC activation (del Peso et al., 1997).

The microinjection of purified PLD or PA into oocytes is able to induce GVBD (Carnero and Lacal, 1993), and PLD and PA signal transduction pathways overlap with the Ras-induced pathway (Carnero and Lacal, 1995). The injection of PLD or PA into *Xenopus* oocytes induces MAPK activation in the absence of *de novo* protein synthesis as well as activating S6 kinase II and MPF. Moreover, the inhibition of DAG production by propanolol does not alter the Ras-induced GVBD. Recently it has been reported that Raf proteins have a phospholi-

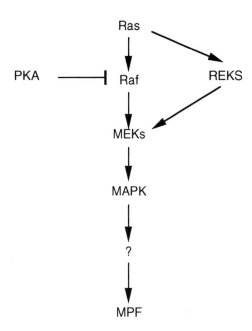

Figure 4 Activation of the Raf/MEK/MAPK cascade by Ras in Xenopus laevis oocytes leads to maturation.

This pathway is blocked by PKA at the level of Raf activation. An alternate pathway for MEK activation is the activation of REKS (Shibuya et al. 1992).

pid binding domain that can bind PA with high affinity but that will not bind other phospholipids as DAG, phosphatidylinositol, sphingosine or ceramide. The inhibition of PA formation significantly reduces both the translocation of Raf to the membrane and its subsequent activation (Ghosh et al., 1996). These results suggest that PA could serve as an important role in Raf activation by linking Raf to the membrane or by directly activating Raf in a manner similar to the way in which DAG activates PKC (Fig. 6).

Finally, the microinjection of Ras proteins or PLD induce the late production of lysophosphatidylcholine as a consequence of phospholipase A2 (PLA2). This effect also seems to be dependent of MAPK activation (Carnero et al., 1994b). The microinjection of purified PLA2 as well as its derived metabolites arachidonic acid or lysoderivates induce maturation of *Xenopus laevis* oocytes (Carnero and Lacal, 1993). However, the inhibition of this enzyme does not block Ras or PLD dependent GVBD (Carnero et al., 1994b), suggesting that PLA2 activity is not needed for Ras-induced GVBD.

5   Remarks

The Ras protein regulates a complex pathway which serves as a critical switch for the control of multiple cellular functions. Its mechanism of activation/inactivation is well known. However, a complete picture of the intricate and complex network of interactions through which Ras controls multiple pathways is still

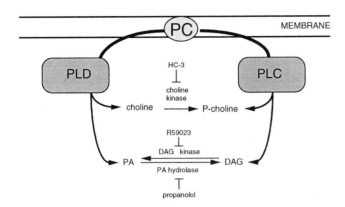

Figure 5 Lipid-derived metabolites by the hydrolysis of phosphatidylcholine (PC) by phospholipase D (PLD) or a putative phospholipase (PLC).

lacking. Ras proteins physically interact with Raf kinase and thus activate the MAP kinase pathway leading to the activation of transcription factors.

Another molecule that interacts with Raf, the Kinase Suppressor of Ras (KSR) has been recently found. KSR is a component of the Ras-dependent signaling pathway that functions to facilitate signal propagation through the MAP kinase cascade (Therrien et al., 1996). KSR is a ceramide activated kinase that associates with and activates Raf at the plasma membrane in a Ras-dependent manner. However, KSR does not phosphorylate Raf (Zhang et al., 1997). The binding to Raf of other molecules such as 14–3–3, p50 or Hsp90 may indicate the existence of a membrane-bound multiprotein signaling complex (Morrison and Cutler, 1997) whose regulation, function and cellular roles require further investigation.

Ras also activates the Jun kinase (JNK)/SAP kinase (SAPK) pathway through Rac/Cdc42. The activation of these G-proteins leads to activation of a MEK kinase (different from Raf) that phosphorylates and activates a different member of the MEK subfamily (MEK 4) which in turn activates the JNK/SAPKs. Activation of these kinases also leads to the activation of transcription factors (see Robinson and Cobb, 1997). How this pathway propagates the Ras signal and how this pathway interacts with the MAP kinase cascade also needs to be investigated.

Two other broad themes also need to be incorporated. One is the involvement of enzymes such as PI3 kinase, phospholipase D or sphingomyelinase (and their derived lipidic metabolites) in the Ras pathways. The other is the damping of the signal. Although extensive efforts are devoted to the Ras activation pathway, subsequent downregulation of the Ras-depending pathways has been largely ignored. Further elucidation of mechanisms by which the Ras signal is downregulated as well as the identification of missing links between the various Ras signalling pathway is required before a clear picture emerges.

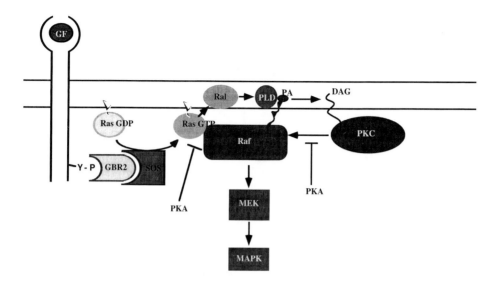

Figure 6   Ras-dependent signal transduction pathway.

Acknowledgements

We thank Greg Hannon, Roberta Maestro and Konstantin Galaktionov for many fruitful discussions and for improving the manuscript with their suggestions. This publication has been possible thanks to the specific support from DGICYT (Grant #PB94–0009) and FIS (Grant #96/2136). AC is the recipient of and EMBO Long-Term Fellowship.

References

Alessi DR, Smythe C, Keyse SM (1993) The human CL100 gene encodes a Tyr/Thr-protein phosphatase which potently and specifically inactivates MAP kinase and suppresses its activation by oncogenic ras in *Xenopus* oocyte extracts. *Oncogene* 8: 2015–2020.

Allende CC, Hinrichs MV, Santos E, Allende JE (1988) Oncogenic ras protein induces meiotic maturation of amphibian oocytes in the presence of protein synthesis inhibitors. *FEBS Lett* 234: 426–430.

Birchmeyer C, Broek D, Wigler M (1985) Ras proteins can induce meiosis in *Xenopus* oocytes. *Cell* 43: 615–621.

Bos JL (1988) The ras gene family and human carcinogenesis. *Mutant Res* 195: 255–271.

Campa MJ, Chang KJ, Molina y Vedia L, Reep BR, Lapetina EG (1991) Inhibition of ras-induced germinal vesicle breakdown in *Xenopus* oocytes by rap-1B. *Biochem Biophys Res Commun* 174: 1–5.

Campa MJ, Glickman JF, Yamamoto K, Chang KJ (1992) The antibiotic azatyrosine suppresses progesterone or [Val12]p21 Ha-ras/insulin-like growth factor I-induced germinal vesicle breakdown and tyrosine phosphorylation of *Xenopus* mitogen-activated protein kinase in oocytes. *Proc Natl Acad Sci USA* 89: 7654–7658.

Carnero A, Lacal, JC (1995) Activation of intracellular kinases in *Xenopus* oocytes by p21ras and phospholipases: a comparative study. *Mol Cell Biol.* 15: 1094–1101.

Carnero A, Lacal, JC (1993) Phospholipase induced maturation of *Xenopus laevis* oocytes. Mitogenic activity of generated metabolites. *J. Cell. Biochem.* 52: 440–448.

Carnero A, Cuadrado A, del Peso L, Lacal JC (1994a) Activation of type D phospholipase by serum stimulation and *ras*-induced transformation in NIH 3T3 cells. *Oncogene* 9: 1387–1395.

Carnero A, Dolfi F, Lacal JC (1994b) Ras-p21 activates phospholipase D and A2, but not phospholipase C or PKC, in *Xenopus laevis* oocytes. *J Cell Biochem* 54: 478–486.

Carnero A, Jimenez B, Lacal JC (1994c) Progesterone but not Ras requires MPF for *in vivo* activation of MAP K and S6 KII. MAPK is an essential conexion point of both signaling pathways. *J. Cell. Biochem.* 55: 465–476.

Chung DL, Brandt-Rauf P, Murphy RB, Nishimura S, Yamaizumi Z, Weinstein IB, Pincus MR (1991) A peptide from the GAP-binding domain of the ras-p21 protein and azatyrosine block ras-induced maturation of *Xenopus* oocytes. *Anticancer Res* 11: 1373–1378.

Cook SJ, MacCormick F (1993) Inhibition by cAMP of Ras-dependent activation of Raf. *Science* 262: 1069–1072.

Daar I, Nebreda AR, Yew N, Sass P, Paules R, Santos E, Wigler M, Vande Woude GF (1991) The ras oncoprotein and M-phase activity. *Science* 253: 74–76.

Daar I, Yew N, Vande Woude GF (1993) Inhibition of mos-induced oocyte maturation by protein kinase A. *J Cell Biol* 120: 1197–1202.

Del Peso L, Hernández R, Esteve P, Lacal JC (1996) Activation of Phospholipase D by Ras proteins is independent of protein kinase C. *J Cell Biochem* 61: 599–608.

Del Peso L, Esteve P, Lacal JC (1997) Activation of Phospholipase D by growth factors and oncogenes in murine fibroblasts follow alternative but cross-talking pathways. *Biochem J* 322: 519–528.

Denhardt DT (1996) Signal-transducing protein phosphorylation cascades mediated by Ras/Rho proteins in the mammalian cell: the potential for multiplex signalling. *Biochem J* 318: 729–747.

Desphande AK, Kung HF (1987) Insulin induction of *Xenopus* laevis oocyte maturation is inhibited by monoclonal antibody against p21 ras proteins. *Mol Cell Biol* 7: 1285–1288.

Fabian JR, Morrison DK, Daar IO (1993) Requirement for Raf and MAP kinase function during the meiotic maturation of *Xenopus* oocytes. *J Cell Biol* 122: 645–652.

Fukuda M, Gotoh Y, Kosako H, Hattori S, Nishida E (1994) Analysis of the ras p21/mitogen-activated protein kinase signalling *in vitro* and in *Xenopus* oocytes. *J Biol Chem* 269: 33097–33101.

Garcia AM, Rowell C, Ackermann K, Kowalczyk JJ, Lewis MD (1993) Peptidomimetic inhibitors of Ras farnesylation and function in whole cells. *J Biol Chem* 268: 18415–18418.

Gibbs JB, Schaber MD, Schofield TL, Scolnick EM, Sigal I (1989) *Xenopus* oocyte germinal-vesicle breakdown induced by vas12Ras is inhibited by a cytosol localized Ras mutant. *Proc Natl Acad Sci USA* 86: 6630–6634.

Gotoh Y, Masuyama N, Dell K, Shirakabe K, Nishida E (1995) Initiation of *Xenopus* oocyte maturation by activation of the mitogen activated protein kinase cascade. *J Biol Chem* 270: 25898–25904.

Graves LM, Bornfeldt KE, Raines EW, Potts BC, MacDonald SG, Ross R, Krebs EG (1993) Protein kinase A antagonizes platelet-derived growth factor-induced signaling by mitogen-activated protein kinase in human arterial smooth muscle cells. *Proc Natl Acad Sci* USA 90: 10300–10304.

Haccard OA, Lewellyn A, Hartley RS, Erikson E, Maller JL (1995) Induction of *Xenopus* oocyte maturation by MAP kinase. *Dev Biol* 168: 677–682.

Hafner S, Adler HS, Mischak H, Janosch P, Heidecker G, Wolfman A, Pippig S, Lohse M, Ueffing M, Kolch W (1994) Mechanism of inhibition of Raf-1 by protein kinase A. *Mol Cell Biol* 14: 6696–6703.

Hagag N, Halegova S, Viola M (1986) Inhibition of growth factor induced differentiation of PC12 cells by microinjection of antibody to ras p21. *Nature.* 319: 680–683.

Hancock JF, Marshall CJ (1993) Posttranslational processing of Ras and Ras-related proteins. In: The Ras Superfamily of GTPases. J.C. Lacal and F. McCormick (eds) CRC Press, New York, USA.

Jiang H, Luo JQ, Urano T, Frankel P, Lu Z, Foster DA, Feig LA (1995) Involvement of Ral GTPase in v-Src-induced phospholipase D activation. *Nature* 378: 409–412.

Johnson AD, Cork RJ, Williams MA, Robinson KR, Smith LD (1990) H-ras(val12) induces cytoplasmic but not nuclear events of the cell cycle. *Cell Regul* 1: 543–554.

Kikuchi A, Williams LT (1996) Regulation of interaction of rasp21 with RalGDS and Raf-1 by cyclic AMP-dependent protein kinase. *J Biol Chem* 271: 588–594.

Korn LJ, Scribel CW, McCormick F, Roth RA (1987) Ras p21 as a potential mediator of insulin action in *Xenopus* oocytes. *Science* 236: 840–843.

Kosako H, Gotoh Y, Nishida E (1994) Requirement for the MAP Kinase kinase/MAP kinase cascade in *Xenopus* oocyte maturation. *EMBO J* 13: 2131–2138.

Koul O, Hauser G (1987) Modulation of rat brain cytosolic phosphatidate phosphohydrolase: effect of cationic amphiphilic drugs and divalent cations. *Arch Biochem Biophys* 253: 453–461.

Koyama S, Chen Y-W, Ikeda M, Muslin AJ, Williams LT, Kikuchi A (1996) Ras-interacting domain of RGL blocks Ras-dependent signal transduction in *Xenopus* oocytes. *FEBS letter* 380: 113–117.

Lacal JC (1990) Diacylglycerol production in *Xenopus laevis* oocytes after microinjection of p21^{ras} proteins is a consequence of activation of phosphatidylcholine metabolism. *Mol Cell Biol* 10: 333–340.

Lacal JC, de la Peña P, Moscat J, García Barreno P, Anderson PS, Aaronson SA (1987a) Rapid stimulation of diacylglycerol production in *Xenopus* oocytes by microinyection of H-ras p21. *Science* 238: 833

Lacal JC, Fleming TP, Warren BS, Blumberg PM, Aaronson SA (1987b) Involvement of functional protein kinase C in the mitogenic response to the H-*ras* oncogene product. *Mol Cell Biol* 7: 4146–4149.

Lacal JC, Moscat J, Aaronson SA (1987c) Novel source of 1,2 diacylglycerol elevated in cells transformed by Ha-Ras oncogene. *Nature* 330: 269–271.

Lacal P, Pennington CY, Lacal JC (1988) Transforming activity of Ras proteins translocated to the plasma membrane by a myristoylation sequence from the *src* gene product. *Oncogene* 2: 533–537.

Leevers SJ, Paterson HF, Marshall CJ (1994) Requirement for Ras in Raf activation is overcome by targeting Raf to the plasma membrane. *Nature* 369: 411–414.

Leon J, Pellicer A (1993) Ras genes involvement in carcinogenesis: lessons from animal model systems. In: The *ras* superfamily of GTPases. JC Lacal and F McCormick (eds) CRC Press, USA.

Losardo JE, Heimer E, Bekesi E, Prinzo K, Scheffler JE, Neri A (1995) Ras-dependent maturation of *Xenopus* oocytes is blocked by modified peptides of GTPase activating protein (GAP). *Int J Pept Protein Res* 45: 194–199.

Macdonald SG, Crews CM, Wu L, Driller J, Clark R, Erikson RL, McCormick F (1993) Reconstitution of the Raf-1-MEK-ERK signal transduction pathway *in vitro*. *Mol Cell Biol* 13: 6615–6620.

Miyake M, MizutaniS, Koide K, Kaziro Y (1996) Unfarnesylated transforming Ras mutant inhibits the Ras-signaling pathway by forming a stable Ras.Raf complex in the cytosol. *FEBS Letter* 378: 15–18.

Morrison DK, Cutler RE (1997) The complexity or Raf-1 regulation. *Curr Op Cell Biol* 9: 174–179.

Nebreda AR, Porras A, Santos E (1993) p21-ras induced meiotic maturation of *Xenopus* oocytes in the absence of protein synthesis: MPF activation is preceeded by activation of MAP ans S6 kinases. *Oncogene* 8: 467–477.

Pomerance M, Schweighoffer F, Tocque B, Pierre M (1992) Stimulation of mitogen-activated protein kinase by oncogenic *ras*-p21 in *Xenopus* oocytes. *J Biol Chem* 267: 16155–16160.

Qiu RG, Chen J, McCormick F, Symonds M (1995) A role for Rho in Ras transformation. *Proc Natl Acad Sci USA* 92: 11781–11785.

Robinson MJ, Cobb MH (1997) Mitogen-activated protein kinase pathways. *Curr Op Cell Biol* 9: 180–186.

Rodriguez-Viciana P, Warne PH, Vanhae-sebroeck B, Waterfield MD, Downward J (1996) Activation of phosphoinositide 3-kinase by interaction with Ras and by point mutation. *EMBO J* 15: 2442–2451.

Sadler SE, Maller JL (1989) A similar pool of cyclic AMP phosphodiesterase in *Xenopus* oocytes is stimulated by insulin, insulin-like growth factor 1 and val12thr59-Ha-ras protein. *J Biol Chem* 264: 856–861.

Sadler SE, Maller JL, Gibbs JB (1990) Transforming Ras proteins acelerate homone induced maturation and stimulate cyclic AMP phosphodiesterase in *Xenopus* oocytes. *Mol Cell Biol* 10: 1689–1696.

Shibuya EK, Ruderman JV (1993) Mos induces the *in vitro* activation of mitogen-activated protein kinases in lysates of frog oocytes and mammalian somatic cells. *Mol Biol Cell* 4: 781–790.

Song J, Pfeffer LM, Foster DA (1991) V-src increases diacylglycerol levels via type D phospholipase-mediated hydrolysis of phosphatidilcholine. *Mol Cell Biol* 11: 4903–4908.

Soto-Cruz I, Magee AI (1995) Effect of synthetic peptides representing the hypervariable region of p21ras on *Xenopus laevis* oocyte maturation. *Biochem J* 306: 11–14.

Stokoe D, Macdonald SG, Cadwallader K, Symons M, Hancock JF (1994) Activation of Raf as a result of recruitment to the plasma membrane. *Science* 264: 1463–1467.

Therrien M, Michaud NR, Rubin GM, Morrison DK (1996) KSR modulates signal propagation within the MAPK cascade. *Genes Dev* 10: 2684–2695.

Wu J, Dent P, Jelinek T, Wolfman A, Weber MJ, Sturgill TW (1993) Inhibition of the EGF-activated MAP kinase signaling pathway by adenosine 3',5'-monophosphate. *Science* 262: 1065–1069.

Zhao J, Kung HF, Manne V (1994) Farnesylation of p21 Ras proteins in *Xenopus* oocytes. *Cell Mol Biol Res* 40: 313–321.

Zhang Y, Yao B, Delikat S, Bayoumy S, Lin XH, Basu S, McGinley M, Chan-Hui PY, Lichenstein H, Kolesnick R (1997) Kinase suppressor of Ras is ceramide-activated protein kinase. *Cell* 89: 63–72.

14 The Multiple Roles of Mos during Meiosis

M.S. Murakami and G.F. Vande Woude

The technique of microinjection has been a critical aspect of experiments which have elucidated the normal function of the Mos oncogene. This technique has also been used to dissect the signal transduction pathway both upstream and downstream of Mos. Antisense oligonucleotides, "prosthetic" oligonucleotides, mRNAs and proteins have been injected into the immature oocytes and cleaving embryos of *Xenopus*. Here, we review the wealth of information which we have collected about this oncogene using the microinjection technique.

The Mos oncogene, which encodes a 39 kDa protein kinase, was originally discovered as the transforming component of the Moloney murine sarcoma virus (Oskarsson et al., 1980). The c-Mos RNA was shown to be highly expressed in germline cells (Propst and Vande Woude, 1985) (Fig. 1), providing the first indication of its normal function. As a consequence of this expression pattern, much work has been done in the oocytes of *Xenopus laevis*, an ideal experimental system because the females have large numbers of oocytes which are easily manipulated (Sagata et al., 1988). In *Xenopus*, the immature stage VI oocytes are arrested in prophase of meiosis I; this arrest is maintained until maturation is initiated when the surrounding follicle cells release the hormone progesterone. Maturation consists of entry into and completion of meiosis I, entry into meiosis II, and arrest at metaphase of meiosis II. Fertilization releases the metaphase II arrest, allows the completion of meiosis II and initiates the series of rapid embryonic cell cycles (Fig. 2). Oocyte maturation was shown to be driven by an activity known as Maturation or M-phase Promoting Factor (MPF) (Fig. 3) and the arrest at metaphase of meiosis II is maintained by an activity known as Cytostatic Factor (CSF) (Fig. 4) (Masui and Markert, 1971). Also, each meiotic cell division (cytokinesis) is remarkably asymmetric, generating one small nonfunctional polar body and one large oocyte. These vast stores of maternal components in the oocyte provide all of the nutrients to the early *Xenopus* embryo, eliminating the need for growth phases during the rapid 30-min cell cycles that follow fertilization. Hence, the meiotic cell cycle is substantially different from the standard somatic cell cycle; it generates a highly specialized cell, which will incorporate the second set of chromosomes from the sperm, then support the series of 12 rapid embryonic cell divisions.

Methods and Tools in Biosciences and Medicine
Microinjection, ed. by J. C. Lacal et al.
© 1999 Birkhäuser Verlag Basel/Switzerland

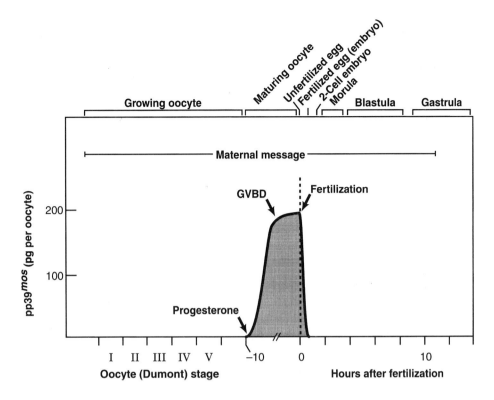

Figure 1   The pattern of Mos expression in *Xenopus laevis*.

The Mos mRNA is a maternal transcript present in oocytes and embryos until the early gastrula stage. The Mos protein is synthesized in response to progesterone and is present at constant levels during the course of oocyte maturation. Fertilization triggers the release from the metaphase II arrest and also triggers the degradation of Mos. The translation of Mos requires the polyadenylation of the mRNA and the degradation of Mos protein requires phosphorylation and ubiqutination.

The role of Mos in many of these unique aspects of meiosis has been examined in detail.

The first breakthrough regarding the normal function of Mos was the observation that the c-mos RNA was abundant in germ cells (Propst and Vande Woude, 1985) (Fig. 1), however, the first direct evidence for the involvement of Mos in meiotic maturation came from the experiment where antisense oligonucleotides directed against *Xenopus* Mos were injected into stage VI oocytes. These antisense oligonucleotides effectively ablated the endogenous mRNA and inhibited the translation of endogenous Mos protein, which in turn, prevented progesterone mediated oocyte maturation (Sagata et al., 1988). Conversely, the injection of *in vitro* transcribed Mos RNA induced oocyte maturation in the absence of progesterone (Sagata et al., 1989). As these experiments implicated Mos as an essential component in the maturation process, it was necessary to determine the relationship of Mos to the activity first described by

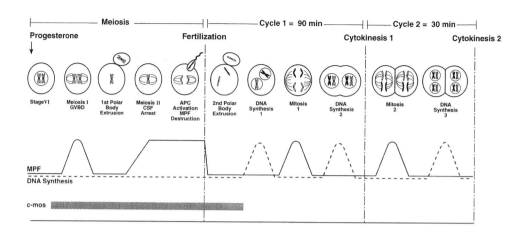

Figure 2   A schematic representation of the events in meiosis and the first two cell cycles.

Oocyte maturation, in response to progesterone, consists of meiosis I (reductive division) and entry into meiosis II. Unfertilized eggs are arrested at metaphase of meiosis II, which is maintained by CSF. This arrest can be mediated by Mos RNA or protein. Fertilization releases the arrest and is followed by the completion of meiosis II (anaphase, telophase, and second polar body extrusion), which requires 20–30 min. The first round of DNA synthesis begins ~30 min after fertilization and takes place in two separate nuclei, which fuse after S-phase. Mitosis of the first cycle begins 60–70 min after fertilization and is completed by 80 min after fertilization. However, cytokinesis continues until 90 min after fertilization. The onset of the second round of DNA synthesis (cycle 2) begins 85 min post fertilization, concurrent with cytokinesis of cycle 1. Thus, the S-phase of the second cell cycle occurs during the cytokinesis of the first cycle resulting in two rounds of DNA replication before the first cytokinesis (see Gerhart, 1980; Graham, 1966; Miake-Lye et al., 1983).

Masui and Markert (Masui and Markert, 1971). This activity, known as MPF (Fig. 3), is a complex of cyclin B and cdc2 (for review see Norbury and Nurse, 1992). Curiously, stage VI oocytes contain pools of inactive pre-MPF (Dunphy and Newport, 1989; Gautier and Maller, 1991) which are complexes of cyclin B and cdc2 where the enzymatic activity is repressed by inhibitory phosphorylations (see below). The activation of this pool of pre-MPF must be triggered by an "initiator", which was postulated to be synthesized prior to MPF activation (Wasserman and Masui, 1975). A careful analysis of endogenous Mos protein expression showed that it was synthesized prior to the activation of MPF, indicating that Mos may be the meiotic "initiator" (Sagata et al., 1989). Subsequently, it was shown that the Mos protein induced oocyte maturation when injected in the presence of cycloheximide (Yew et al., 1992). These findings implicated Mos as the meiotic "initiator", however meiotic maturation can be blocked by the injection of a kinase defective cdc2, indicating these two kinases (Mos and cdc2) act synergistically in this process (Nebreda et al., 1995).

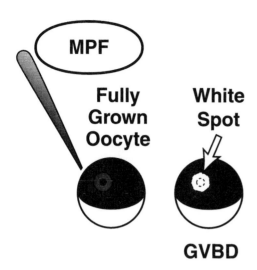

Figure 3   MPF assay.

The original assay for MPF was described by Masui and Markert in 1971. Cytoplasmic lysates were taken from mature eggs and injected into immature stage VI oocytes. The activity in the lysates (termed MPF for Maturation Promoting Factor) initiated all the events involved in oocyte maturation. In the immature stage VI oocyte the germinal vesicle (nucleus) is situated toward the center of the oocyte. Before the nuclear envelope breaks down, the germinal vesicle migrates to the surface of the oocyte, displacing the pigment, creating a white spot in the dark animal pole. Thus, Germinal Vesicle Break Down (GVBD) is generally scored by the appearance of a white spot in the animal pole of the oocyte.

In addition to triggering the initial oocyte maturation process in *Xenopus*, Mos has also been shown to be necessary at all stages of *Xenopus* oocyte maturation. The haploid oocyte is the end product of the meiotic cell cycle which consists of two consecutive M-phases without an intervening S-phase. In *Xenopus*, the oocytes become competent for DNA replication after meiosis I, therefore DNA replication must be actively repressed during meiosis (Benbow and Ford, 1975; Gurdon, 1967). Oocytes will undergo DNA replication if they are injected with antibodies which inhibit Mos or if they are placed in cycloheximide 30 min after Germinal Vesicle Breakdown (GVBD) (Furuno et al., 1994). The ablation of Mos protein by the injection of antisense oligonucleotides after GVBD also prevents entry into meiosis II (Daar et al., 1991; Kanki and Donoghue, 1991). While these experiments indicate that Mos is required to inhibit DNA synthesis between meiosis I and meiosis II, and that it is required for entry into meiosis II, it is also apparent that other proteins or activities must also be involved. The injection of Mos protein does not inhibit DNA replication in oocytes treated with cycloheximide or in eggs (Furuno et al., 1994; Murakami and Vande Woude, 1998) and oocytes injected with Mos protein in the presence of cycloheximide complete meiosis I but do not progress to meiosis II (Yew et al., 1992). The analysis of oocytes derived from Mos -/- mice has revealed that Mos is also required for the proper assembly of the meiotic spindle, and may regulate the highly asymmetric cytokinesis which generates the small non-functional polar bodies. The microtubule structures in the Mos -/- oocytes were highly irregular and a large number of oocytes generated an abnormally large polar body (Choi et al., 1996; Verlhac et al., 1996). Furthermore, the polar bodies from these oocytes persist rather than degrade and some of the oocytes appear to undergo a third meiotic division (Choi et al., 1996; Verlhac et al., 1996).

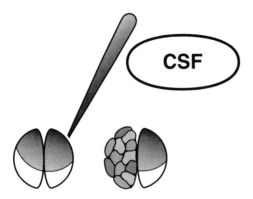

Figure 4 CSF assay.

The original assay for CSF was also de-scribed by Masui and Markert. In this assay the eggs are fertilized and allowed to pro-gress to the two-cell stage. At this point, cy-toplasmic lysates are injected into one of the two blastomeres. If the lysate possesses CSF activity, the injected cell will arrest at meta-phase, while the uninjected cell continues to divide. Proteins which have been shown to possess CSF activity include Mos, Ras, Raf, Mek, and MAPK.

One of the most critical functions of Mos is the maintenance of the arrest at metaphase of meiosis II (Sagata, 1996; Sagata et al., 1989). Cytostatic Factor (CSF) is the activity that maintains the metaphase II arrest in the egg. CSF was discovered upon the injection of unfertilized egg extract into one blastomere of a two-cell embryo (Masui and Markert, 1971); the injected blastomere arrests at metaphase while the uninjected blastomere continues to divide (Fig. 4). The observation that injection of Mos RNA could also induce CSF arrest (in the blas-tomere assay) implicated Mos as a critical component of this activity (Sagata et al., 1989). Moreover, the immunodepletion of Mos from a CSF extract resulted in the loss of CSF activity (Sagata et al., 1989), and the oocytes from Mos -/- mice do not arrest at meiosis II (Colledge et al., 1994; Hashimoto et al., 1994). One characteristic of CSF is its disappearance after fertilization in a calcium-dependent manner (Masui and Markert, 1971). Likewise, endogenous Mos pro-tein is degraded with the same kinetics as CSF. Interestingly, the degradation of Mos/CSF follows that of cyclin B/MPF by 10–15 min (Watanabe et al., 1991). It is possible that this delayed degradation of Mos may serve to lengthen the first mitotic cell cycle; the injection of Mos RNA and Mos protein into eggs and embryos delays entry into the mitotic M-phase (Murakami and Vande Woude, 1998). While it is clear that Mos has CSF activity, other components are essen-tial to this process, as high levels of Mos protein at meiosis I do not result in M-phase arrest. One such component may be cdk2; the injection of antisense oli-gonucleotides directed against cdk2 abolished the ability of oocytes to arrest at meiosis II (Gabrielli et al., 1993). However, it has also been shown that the inhi-bition of cdk2 by the injection of a cdk2 inhibitor, p21/CIP, had no effect on CSF arrest (Furuno et al., 1997). Thus, the exact biochemical components which result in CSF arrest are still undefined, but is likely to involve the inhibition of Anaphase Promoting Complex, the activity which mediates the degradation of cyclin (for review, see King et al., 1994). It is interesting to note that the activa-tion of the mitotic checkpoint, which occurs in response to mitotic spindle per-turbation, also arrests the cell cycle in M-phase. This M-phase arrest is similar, in some regards, to the arrest observed at metaphase of meiosis II (CSF arrest).

Mitotic checkpoint arrests have been reconstituted *in vitro* by adding sperm nuclei and nocodozol to cycling *Xenopus* extracts (Minshull et al., 1994). This mitotic checkpoint requires the activation of MAPK and can be ablated by adding a MAPK phosphatase (Minshull et al., 1994). The mitotic checkpoint can also be ablated with antibodies raised against *Xenopus* MAD2, a gene initially isolated from an anti-microtubule drug screen in yeast (Chen et al., 1996; Li and Murray, 1991; Li and Benezra, 1996). The addition of Mos to *Xenopus* extracts also results in a MAPK-dependent M-phase arrest. However, the Mos mediated arrest does not depend on the addition of sperm nuclei or nocodozol and cannot be ablated by anti-MAD2 antibodies (Chen et al., 1996; Minshull et al., 1994). Thus, the mitotic checkpoint arrest and the CSF arrest at meiosis II both involve the activation of MAPK, however, the precise biochemical differences between these two types of M-phase arrests have not yet been defined.

Because Mos is a critical component of the maturation process, it follows that the regulation of Mos translation would also be a highly regulated process. In fact, *Xenopus* oocytes contain large pools of mRNA (Fig. 1); transcription is not required for oocyte maturation or the rapid cell cycles that follow fertilization (for reviews see Bachvarova, 1992; Gerhart, 1980). During this time protein expression is regulated at the level of translation and degradation (for review see Howe et al., 1995; Richter, 1996; Wickens et al., 1996). Polyadenylation enhances the rate of translation, and both cyclin B and Mos mRNAs are polyadenylated in response to progesterone (Sheets et al., 1994). Polyadenylation is regulated by elements in the 3' untranslated region (UTR): the highly conserved poly(A) signal -AAUAAA and a U-rich region known as the cytoplasmic polyadenylation element (CPE) (for reviews see Richter, 1996; Wickens et al., 1996). The 3' UTRs of cyclin B and Mos are sufficient to confer both the polyadenylation and the translation pattern of these genes to a heterologous reporter gene (Sheets et al., 1994). Furthermore, oocyte maturation in response to progesterone can be prevented by ablating the polyadenylation elements in the 3' UTR of Mos with an antisense oligonucleotide. Mos mRNA polyadenylation and oocyte maturation are restored upon injection of a prosthetic RNA, which hybridizes with the truncated Mos mRNA and restores the 3' UTR (Sheets et al., 1995). In addition, the polyadenylation of Mos mRNA is required for the polyadenylation of other mRNAs such as cyclin B (Ballantyne et al., 1997). These experiments have focused current studies on the regulation of cytoplasmic polyadenylation as well as other mechanisms of translational control. Toward this end, one protein that binds to the CPE has been identified and has been shown to be necessary for both cytoplasmic polyadenylation and progesterone-mediated oocyte maturation (Hake and Richter, 1994; Stebbins-Boaz et al., 1996).

The hormone progesterone initiates a signal transduction cascade which ultimately leads to meiotic resumption and MPF activation. While the biochemical mechanism of progesterone signaling is unknown, progesterone treatment results in a transient reduction in the intracellular levels of cyclic AMP (cAMP) (Fig. 5; for review see Smith, 1989). This, in turn, leads to a decrease in cAMP-

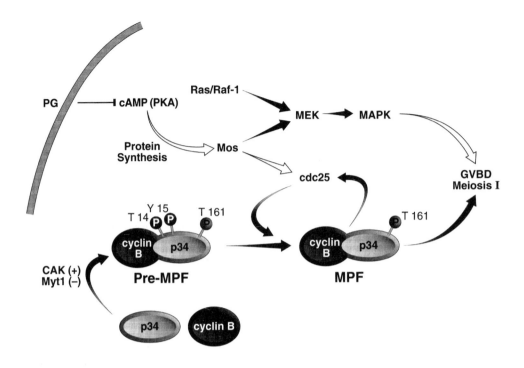

Figure 5   Model of the events which follow progesterone addition to oocytes.

Progesterone addition to oocytes decreases the activity of PKA, which in turn increases the translation of Mos protein. Mos then phosphorylates and activates MEK which phosphorylates and activates MAPK. The addition of progesterone to oocytes also stimulates the translation of other proteins which, together with Mos, result in the activation of MPF. In immature stage VI oocytes, MPF is kept in a latent inactive form by inhibitory phosphorylations on threonine 14 and tyrosine 15. The activation of this pool of inactive pre-MPF is mediated by the dephosphorylation of threonine 14 and tyrosine 15 by the cdc25 phosphatase.

dependent kinase (PKA) activity. The injection of PKA catalytic subunit blocks progesterone mediated oocyte maturation (Maller and Krebs, 1977). Conversely, the inhibition of PKA by the injection of a PKA inhibitor peptide or the regulatory subunit of PKA induces oocyte maturation in the absence of progesterone (Maller and Krebs, 1977). Thus elevated levels of PKA activity serve to inhibit maturation, and the down regulation of this activity is one of the first steps in the maturation process. In order to elucidate the downstream effects of PKA downregulation, the effects of PKA on Mos function and Mos protein synthesis have been examined. The injection of PKA catalytic subunit inhibited the synthesis of endogenous Mos protein in response to progesterone (Matten et al., 1994) and oocyte maturation resulting from the injection of PKA regulatory subunit was shown to be dependent on endogenous Mos product (Daar et al., 1993). These results indicate that some of the inhibitory effects of PKA are mediated through the regulation of Mos protein synthesis. Elevated levels of

PKA activity also prevent the activation of MPF by inhibiting cdc2 tyrosine de-phosphorylation (Matten et al., 1994; Rime et al., 1992). Thus, PKA acts at multiple points to inhibit the maturation process and is a potent antagonist of the Mos/MAPK pathway (Fig. 5).

A critical downstream target of Mos is the Mitogen Activated Protein Kinase (MAPK). The activation of Mos leads to the activation of MAPK in *Xenopus* oocytes and oocyte extracts (Nebreda and Hunt, 1993; Posada et al., 1993; Shibuya and Ruderman, 1993). The activation of MAPK is mediated by the direct phosphorylation of and subsequent activation of MAPK kinase (MEK) by Mos (Pham et al., 1995; Posada et al., 1993; Resing et al., 1995). The meiotic cell cycle is unique in that MAPK activity is high throughout the lengthy period of meiosis (Choi et al., 1996; Ferrell et al., 1991; Posada and Cooper, 1992; Verlhac et al., 1994); by comparison, somatic cell cycles typically have a transient peak of activity in response to growth factor stimulation (for review, see Kosako et al., 1994). The prolonged activation of MAPK throughout the meiotic cell cycle appears to be a programmed response in the oocyte, a direct result of the expression pattern of Mos. The activation of MAP kinase is also enhanced by the fact that the oocytes possess a positive feedback loop between Mos and MAP kinase (Matten et al., 1996; Roy et al., 1996). In response to progesterone, MAP kinase and MPF are activated simultaneously (Ferrell et al., 1991; Gotoh et al., 1991; Posada and Cooper, 1992); however, the function of MAPK in the signaling cascade leading to oocyte maturation is only partially understood. The inhibition of MAP kinase with the CL100 phosphatase or an antibody that inhibits MEK prevents progesterone induced oocyte maturation (Gotoh et al., 1995; Kosako et al., 1994), and the injection of thiophosphorylated MAP kinase or activated MEK induces oocyte maturation in the absence of progesterone (Haccard et al., 1995; Huang et al., 1995). Collectively, these experiments suggest that the activation of MAP kinase is essential for oocyte maturation. However, GVBD can occur without MAP kinase activation under experimental conditions where the activity of MAP kinase is inhibited by either a dominant-negative Raf or a dominant-negative KSR (Kinase Suppressor of Ras) (Fabian et al., 1993) (Michaud, Thierren, Morrison, personal communication, 1996). Similarly, the injection of a nondestructible cyclin protein in the presence of elevated PKA activity results in the activation of MPF, but not MAP kinase (Rime et al., 1992). Conversely, the injection of Mos protein in the presence of elevated PKA activity results in the activation of MAPK but not MPF (Matten et al., 1994). The injection of Mos protein into intact oocytes leads to the rapid activation of MAP kinase hours before the activation of MPF (Posada et al., 1993), and the addition of Mos protein to oocyte lysates results in the activation of MAP kinase but not MPF (Nebreda and Hunt, 1993; Shibuya and Ruderman, 1993). And the addition of exogenous MAP kinase to oocyte extracts restores the ability of Mos to activate MPF (Huang and Ferrell, 1996). While these experiments demonstrate that MAP kinase and MPF can be activated independently, optimally both kinases are required for normal maturation. At the cytological level, the injection of nondestructible cyclin in the presence of cycloheximide (which results in the activation of MPF but not MAPK) results in GVBD, but the first meiotic spindle fails to form (Huchon et al., 1993; Rime et al., 1992) (Duesbery; unpublished). In mouse oocytes, the activation of MAPK in the absence of MPF results in partial chromosome condensation and the formation of microtubule arrays (Choi et al., 1996). The activation of MPF in the absence of MAPK results in abnormal microtubule organization and large, persistent polar bodies (Choi et al., 1996; Verlhac et al., 1996). Thus, the elevated MAPK activity observed during oocyte maturation is postulated to maintain chromosome condensation between meiosis I and meiosis II and is thought to be important for the formation of the meiotic spindle (Choi et al., 1996; Choi et al., 1996; Verlhac et al., 1994). Currently, we are interested in the substrates which lie downstream of MAPK and we have recently shown that an epitope of CENP-E, a kinesin-like motor protein, is masked at meiosis II in a Mos-dependent manner (Duesbery et al., 1997). The masking of this epitope may reflect differences between the two meiotic spindles, and may implicate CENP-E as a downstream target of Mos/MAPK.

Interestingly, MAP kinase activity is not detected to the same extent in the M-phases of the somatic cell cycle or the early embryonic cell cycles of *Xenopus* (Ferrell et al., 1991; Hartley et al., 1994). However, the activation of MAP kinase in the early embryonic cell cycles by Mos, Ras or thiophosphorylated MAPK results in CSF arrest (Choi et al., 1996; Daar et al., 1991; Haccard et al., 1993; Posada et al., 1993). These experiments indicate that the activation of MAP kinase is sufficient to induce CSF arrest in the biochemical context of meiosis II or the early embryonic cycles (but not meiosis I). Furthermore, it is clear that the timing of MAPK activation in the cell cycle is also critical. The activation of MAPK during the early portion of the *mitotic* cell cycle results in a G2/M delay or arrest (Abrieu et al., 1997; Murakami and Vande Woude, 1998; Walter et al., 1997; Shibuya, 1998), while the activation of MAPK later in the mitotic cycle leads to CSF arrest (Abrieu et al., 1996).

Although we have learned a great deal about the normal function of the Mos oncogene, we know very little about the substrates which lie downstream of the Mos/MAPK pathway. MPF, the activity first described by Masui (1971) and Smith (1971) is regulated by the synthesis and degradation of cyclin. MPF is also regulated by phosphorylation on three critical residues: threonine 14, tyrosine 15, and threonine 161 (Fig. 5). The phosphorylation on threonine 161 is required for full activation of the kinase; however, the activity of the kinase that mediates this phosphorylation, MO15, does not appear to be regulated during the course of oocyte maturation (Brown et al., 1994). In stage VI oocytes the pre-MPF complex of cyclin B and cdc2 is maintained in an inactive state by inhibitory phosphorylations on tyrosine 15 and threonine 14 (Dunphy and Newport, 1989; Gautier and Maller, 1991). In a mitotic extract, these phosphorylations are mediated by the Wee1 and Myt1 kinases (see Dunphy, 1994; Mueller et al., 1995). However, Wee1 is not present in the stage VI oocyte, therefore the inhibitory phosphorylation of pre-MPF may be mediated only by Myt1 (Murakami and Vande Woude, 1998). During oocyte maturation, the dephosphorylation of these sites is mediated by the Cdc25 phosphatase which, in turn, is also activated by phosphorylation (Dunphy, 1994). We have seen that the injection of Mos protein into stage VI oocytes results in the tyrosine dephosphorylation of cdc2 and the hyperphosphorylation of cdc25, but cdc25 does not appr'ar to be a direct functional target of Mos/MAPK (Kuang and Ashorn, 1993; Kuang et al., 1994; Murakami and Vande Woude, 1997). Alternatively, Mos may act to inactivate the Myt1 or other upstream components such as Polo kinase or Chk1 (Furnari et al., 1997; Kumagai and Dunphy, 1995; Peng et al., 1997; Sanchez et al., 1997).

In summary, the technique of microinjection has been an essential component of the experiments which have elucidated the normal function of the Mos oncogene. The injection of antisense oligonucelotides demonstrated that Mos was involved in oocyte maturation and the injection of Mos mRNA into cleaving embryos demonstrated that Mos was a component of CSF. The injection of prosthetic oligonucleotides was used to demonstrate the importance of Mos polyadenylation in the maturation process and the injection of the regulatory and catalytic subunits of PKA illustrated that some of the inhibitory effects of this enzyme are mediated at the level of Mos synthesis. The activation of MAPK by Mos was first shown following the injection of Mos protein into stage VI oocytes, and the observation that the injection of an activated form of MAPK mediated both oocyte maturation and CSF arrest confirmed the importance of MAPK as a downstream target of Mos. Thus, the microinjection technique has been and continues to be invaluable as we dissect the detailed molecular mechanisms of Mos function.

References

Abrieu A, Fisher D, Simon M-N et al. (1997) MAPK inactivation is required for the G_2 to M-phase transition of the first mitotic cell cycle. EMBO J 16: 6407–6413.

Abrieu A, Lorca T, Labbe J-C et al. (1996) MAP kinase does not inactivate, but rather prevents the cyclin degradation pathway from being turned on in Xenopus egg extracts. J Cell Sci 109: 239–246.

Bachvarova RF (1992) A maternal tail of poly(a): The long and the short of it. Cell 69: 895–897.

Ballantyne S, Daniel DL, Wickens M (1997) A dependent pathway of cytoplasmic polyadenylation reactions linked to cell cycle control by c-mos and CDK1 activation. Mol Bio Cell 8: 1633–1648.

Benbow RM and Ford CC (1975) Cytoplasmic control of nuclear DNA synthesis during early development of Xenopus laevis: A cell-free assay. Proc. Natl. Acad. Sci. USA 72, 2437–2441.

Brown AJ, Jones T, Shuttleworth J (1994) Expression and activity of $p40^{mo15}$, the catalytic subunit of cdk-activating kinase, during Xenopous oogenesis and embryogenesis. Mol Biol Cell 5: 921–932.

Chen R-H, Waters JC, Salmon ED et al. (1996) Association of spindle assembly checkpoint component XMAD2 with unattached kinetochores. Science 274: 242–246.

Choi T, Fukasawa K, Zhou R et al. (1996) The Mos/mitogen-activated protein kinase (MAPK) pathway regulates the size and degradation of the first polar body in maturing mouse oocytes. Proc Natl Acad Sci USA 93: 7032–7035.

Choi T, Rulong S, Resau J et al. (1996) Mos/mitogen-activated protein kinase can induce early meiotic phenotypes in the absence of maturation-promoting factor: A novel system for analyzing spindle formation during meiosis I. Proc Natl Acad Sci USA 93: 4730–4735.

Colledge WH, Carlton MBL, Udy GB et al. (1994) Disruption of c-*mos* causes parthenogenetic development of unfertilized mouse eggs. Nature *370*: 65–68.

Daar I, Nebreda AR, Yew N et al. (1991) The *ras* oncoprotein and M-phase activity. Science *253*: 74–76.

Daar I, Paules RS, Vande Woude GF (1991) A characterization of cytostatic factor activity from *Xenopus* eggs and c-mos-transformed cells. J Cell Biol *114*: 329–335.

Daar I, Yew N, Vande Woude GF (1993) Inhibition of mos-induced oocyte maturation by protein kinase A. J Cell Biol *120*: 1197–1202.

Duesbery NS, Choi T, Brown KD, et al. (1997) CENP-E is an essential kinetochore motor in maturing oocytes and is masked during Mos- dependent, cell cycle arrest at metaphase II. Proc Natl Acad Sci USA *94*: 9165–9170.

Dunphy WG (1994) The decision to enter mitosis. Trends in Cell Biol *4*: 202–207.

Dunphy WG and Newport JW (1989) Fission yeast p13 blocks mitotic activation and tyrosine dephosphorylation of the *Xenopus* cdc2 protein kinase. Cell *58*: 181–191.

Fabian JR, Morrison DK, Daar IO (1993) Requirement for Raf and MAP kinase function during the meiotic maturation of *Xenopus* oocytes. J Cell Biol *122*: 645–652.

Ferrell JE, Wu M, Gerhart JC et al. (1991) Cell cycle tyrosine phosphorylation of p34^{cdc2} and a microtubule-associated protein kinase homolog in *Xenopus* oocytes and eggs. Mol Cell Biol *11*: 1965–1971.

Furnari B, Rhind N, Russell P (1997) Cdc25 mitotic inducer targeted by Chk1 DNA damage checkpoint kinase. Science *277*: 1495–1497.

Furuno N, Nishizawa M, Okazaki K et al. (1994) Suppression of DNA replication via Mos function during meiotic divisions in *Xenopus* oocytes. EMBO J *13*: 2399–2410.

Furuno N, Ogawa Y, Iwashita J et al. (1997) Meiotic cell cycle in *Xenopus* oocytes is independent of cdk2 kinase. EMBO J *16*: 3860–3866.

Gabrielli BG, Roy LM, Maller JL (1993) Requirement for Cdk2 in cytostatic factor-mediated Metaphase II arrest. Science *259*: 1766–1769.

Gautier J and Maller JL (1991) Cyclin B in *Xenopus* oocytes: implications for the mechanism of pre-MPF activation. EMBO J *10*: 177–182.

Gerhart JG (1980) Mechanisms regulating pattern formation in the amphibian egg and early embryo. In Biological Regulation and Development, R. F. Goldberger, ed. (New York: Plenum Publishing), pp. 133–316.

Gotoh Y, Masuyama N, Dell K et al. (1995) Initiation of *Xenopus* oocyte maturation by activation of the mitogen-activated protein kinase cascade. J Biol Chem *270*: 25898–25904.

Gotoh Y, Nishida E, Matsuda S et al. (1991) *In vitro* effects on microtubule dynamics of purified *Xenopus* M-phase-activated dynamics of purified *Xenopus* M-phase-activated MAP kinase. Nature *349*: 251–254.

Graham CF (1966) The regulation of DNA synthesis and mitosis in multinucleate frog eggs. J Cell Sci *1*: 363–374.

Gurdon JB (1967) On the origin and persistence of a cytoplasmic state inducing nuclear DNA synthesis in frogs' eggs. Proc Natl Acad Sci USA *58*: 545–552.

Haccard O, Lewellyn A, Hartley RS et al. (1995) Induction of *Xenopus* oocyte meiotic maturation by MAP kinase. Dev Biol *168*: 677–682.

Haccard O, Sarcevic B, Lewellyn A et al. (1993) Induction of metaphase arrest in cleaving *Xenopus* embryos by MAP kinase. Science *262*: 1262–1265.

Hake LE and Richter JD (1994) CPEB is a specificity factor that mediates cytoplasmic polyadenylation during *Xenopus* oocyte maturation. Cell *79*: 617–627.

Hartley RS, Lewellyn AL, Maller JL (1994) MAP kinase is activated during mesoderm induction in *Xenopus* laevis. Dev Biol *163*: 521–524.

Hashimoto N, Watanabe N, Furuta Y et al. (1994) Parthenogenetic activation of oocytes in c-*mos*-deficient mice. Nature *370*: 68–71.

Howe JA, Howell M, Hunt T et al. (1995) Identification of a developmental timer regulating the stability of embryonic cyclin A and a new somatic A-type cyclin at gastrulation. Genes & Dev *9*: 1164–1176.

Huang C-YF and Ferrell JEJ (1996) Dependence of Mos-induced Cdc2 activation on MAP kinase function in a cell-free system. EMBO J. *15*, 2169–2173.

Huang W, Kessler DS, Erikson RL (1995) Biochemical and biological analysis of Mek1 phosphorylation site mutants. Mol. Biol. Cell. *6*, 237–245.

Huchon D, Rime H, Jessus C et al. (1993) Control of metaphase I formation in *Xenopus* oocyte: Effects of an indestructible cyclin B and of protein synthesis. Biol Cell *77*: 133–141.

Kanki JP and Donoghue DJ (1991) Progression from meiosis I to meiosis II in *Xenopus* oocytes requires *de novo* translation of the *mos^{xe}* protooncogene. Proc Natl Acad Sci USA *88*: 5794–5798.

King RW, Jackson PK, Kirschner MW (1994) Mitosis in transition. Cell *79*: 563–571.

Kosako H, Gotoh Y, Nishida E (1994) Regulation and function of the MAP kinase cascade in *Xenopus* oocytes. J Cell Sci Supp *18*: 115–119.

Kosako H, Gotoh Y, Nishida E (1994) Requirement for the MAP kinase kinase/MAP kinase cascade in *Xenopus* oocyte maturation. EMBO J *13*: 2131–2138.

Kuang J and Ashorn CL (1993) At least two kinases phosphorylate the MPM-2 epitope during *Xenopus* oocyte maturation. J Cell Biol *123*: 859–868.

Kuang J, Ashorn CL, Gonzalez-Kuyvenhoven M et al. (1994) cdc25 is one of the MPM-2 antigens involved in the activation of maturation-promoting factor. Mol Biol Cell *5*: 135–145.

Kumagai A and Dunphy WG (1995) Control of the Cdc2/cyclin B complex in *Xenopus* egg extracts arrested at a G2/M checkpoint with DNA synthesis inhibitors. Mol Biol Cell *6*: 199–213.

Li R and Murray AW (1991) Feedback Control of Mitosis in Budding Yeast. Cell *66*: 519–531.

Li Y and Benezra R (1996) Identification of a Human Mitotic Checkpoint Gene: hsMAD2. Science *274*: 246–248.

Maller JL and Krebs EG (1977) Progesterone-stimulated meiotic cell division in *Xenopus* oocytes. J Biol Chem *252*: 1712–1718.

Masui Y and Markert CL (1971) Cytoplasmic control of nuclear behavior during meiotic maturation of frog oocytes. J Exp Zool *177*: 129–146.

Matten W, Daar I, Vande Woude GF (1994) Protein kinase A acts at multiple points to inhibit *Xenopus* oocyte maturation. Mol Cell Biol *14*: 4419–4426.

Matten WT, Copeland TD, Ahn NG et al. (1996) Positive feedback between MAP Kinase and Mos during *Xenopus* oocyte maturation. Dev Biol *179*: 485–492.

Miake-Lye R, Newport J, Kirschner M (1983) Maturation-promoting factor induces nuclear envelope breakdown in cyclo-heximide-arrested embryos of *Xenopus laevis*. J Cell Biol *97*:81–91.

Minshull J, Sun H, Tonks NK et al. (1994) A MAP kinase-dependent spindle assembly checkpoint in *Xenopus* egg extracts. Cell *79*: 475–486.

Mueller PR, Coleman TR, Kumagai A et al. (1995) Myt1: A membrane-associated inhibitory kinase that phosphorylates Cdc2 on both threonine-14 and tyrosine-15. Science *270*: 86–90.

Murakami MS and Vande Woude GF (1998) Analysis of the early embryonic cell cycles of *Xenopus*; regulation of cell cycle length by Xe-wee1 and Mos. Development *125*: 237–248.

Murakami MS and Vande Woude GF (1997) Mechanisms of *Xenopus* oocyte maturation. Methods Enzymol *283*: 584–600.

Nebreda AR, Gannon JV, Hunt T (1995) Newly synthesized protein(s) must associate with p35^{cdc2} to activate MAP kinase and MPF during progesterone-induced. EMBO J *14*: 5597–5607.

Nebreda AR and Hunt T (1993) The Mos protooncogene protein kinase turns on and maintains the activity of MAP kinase, but not MPF, in cell-free extracts of *Xenopus* oocytes and eggs. EMBO J *12*: 1979–1986.

Norbury C and Nurse P (1992) Animal cell cycles and their control. Annu Rev Biochem *61*: 441–470.

Oskarsson M, McClements W, Blair DG et al. (1980) Properties of a normal mouse cell DNA sequence (sarc) homologous to the *src* sequence of Moloney sarcoma virus. Science *207*: 1222–1224.

Peng C-Y, Graves PR, Thoma RS et al. (1997) Mitotic and G2 checkpoint control: regulation of 14–3-3 protein binding by phosphorylation of Cdc25c on Serine-216. Science *277*: 1501–1505.

Pham CD, Arlinghaus RB, Zheng CF et al. (1995) Characterization of MEK1 phosphorylation by the v-Mos protein. Oncogene *10*: 1683–1688.

Posada J and Cooper JA (1992) Requirements for phosphorylation of MAP kinase during meiosis in *Xenopus* oocytes. Science *255*: 212–215.

Posada J, Yew N, Ahn NG et al. (1993) Mos stimulates MAP kinase in *Xenopus* oocytes and activates a MAP kinase kinase *in vitro*. Mol Cell Biol *13*: 2546–2553.

Propst F and Vande Woude GF (1985) c-*mos* proto-oncogene transcripts are expressed in mouse tissues. Nature *315*: 516–518.

Resing KA, Mansour SJ, Hermann AS et al. (1995) Determination of v-mos-catalyzed phosphorylation sites and autophosphorylation sites on MAP kinase kinase by ESI/MS. Biochemistry *34*: 2610–2620.

Richter JD (1996) Dynamics of Poly(A) addition and removal during development. In: Translational Control, J.W.B. Hershey, M.B. Mathews and N. Sonenberg, eds. (Cold Spring Harbor, NY: Cold Spring Harbor Laboratory Press), pp. 481–503.

Rime H, Haccard O, Ozon R (1992) Activation of P34^{cdc} kinase by cyclin is negatively regulated by cyclic amp-dependent protein kinase in *Xenopus* oocytes. Dev Biol *151*: 105–110.

Roy LM, Haccard O, Izumi T, et al. (1996) Mos proto-oncogene function during oocyte maturation in *Xenopus*. Oncogene *12*, 2203–2211.

Sagata N (1996) Meiotic metaphase arrest in animal oocytes: its mechanisms and biological significance. Trends in Cell Biol *6*: 22–28.

Sagata N, Daar I, Oskarsson M, et al. (1989) The product of the *mos* proto-oncogene product as a candidate "initiator" for oocyte maturation. Science *245*: 643–646.

Sagata N, Oskarsson M, Copeland T, et al. (1988) Function of c-*mos* proto-oncogene product in meiotic maturation in *Xenopus* oocytes. Nature *335*: 519–525.

Sagata N, Watanabe N, Vande Woude GF et al. (1989) The c-*mos* proto-oncogene product is a cytostatic factor responsible for meiotic arrest in vertebrate eggs. Nature *342*: 512–518.

Sanchez Y, Wong C, Thoma RS et al. (1997) Conservation of the Chk1 checkpoint pathway in mammals: linkage of DNA damage to Cdk regulation through Cdc25. Science *277*: 1497–1501.

Sheets MD, Fox CA, Hunt T et al. (1994) The 3'-untranslated regons of c-*mos* and cyclin mRNAs stimulate translation by regulating cytoplasmic polyadenylation. Genes & Dev. *8*: 926–938.

Sheets MD, Wu M, Wickens M (1995) Polyadenylation of c-mos mRNA as a control point in *Xenopus* meiotic maturation. Nature *374*: 511–516.

Shibuya EK and Ruderman JV (1993) Mos induces the *in vitro* activation of mitogen-activated protein kinases in lysates of frog oocytes and mammalian somatic cells. Mol Biol Cell *4*: 781–790.

Shibuya EK (1999) in press.

Smith LD (1989) The induction of oocyte maturation: transmembrane signalling events and regulation of the cell cycle. Development *107*: 685–699.

Smith LD and Ecker RE (1971) The interaction of steroids with Rana pipiens oocytes in the induction of maturation. Dev Biol *25*: 233–247.

Stebbins-Boaz B, Hake LE, Richter JD (1996) CPEB controls the cytoplasmic polyadenylaton of cyclin Cdk2 and c-*mos* mRNAs and is necessary for oocyte maturation in *Xenopus*. EMBO J *15*: 2582–2592.

Verlhac M-H, Kubiak JZ, Clarke HJ et al. (1994) Microtubule and chromatin behavior follow MAP kinase activity but not MPF activity during meiosis in mouse oocytes. Development *120*: 1017–1025.

Verlhac M-H, Kubiak JZ, Weber M et al. (1996) Mos is required for MAP kinase activation and is involved in microtubule organization during meiotic maturation in the mouse. Development *122*: 815–822.

Walter SA, Guadagno TM, Ferrell JE (1997) Induction of a G$_2$-Phase Arrest in *Xenopus* Egg Extracts by Activation of p42 Mitogen-activated Protein Kinase. Mol Biol Cell *8*: 2157–2169.

Wasserman WJ and Masui Y (1975) Effects of cycloheximide on a cytoplasmic factor initiating meiotic maturation in *Xenopus laevis* oocytes. Exp Cell Res *91*: 381–388.

Watanabe N, Hunt T, Ikawa Y et al. (1991) Independent inactivation of MPF and cytostatic factor (Mos) upon fertilization of *Xenopus* eggs. Nature *352*: 247–248.

Wickens M, Kimble J, Strickland S (1996) Translational control of developmental decisions. In: Translational Control, J.W.B. Hershey, M.B. Matthews and N. Sonenberg, eds. (Cold Spring Harbor, NY: Cold Spring Harbor Laboratory Press), pp. 411–450.

Yew N, Mellini ML, Vande Woude GF (1992) Meiotic initiation by the *mos* protein in *Xenopus*. Nature *355*: 649–652.

15 Use of the *Xenopus* Oocyte System to Study RNA Transport

L.A. Allison

Contents

1   Introduction

Xenopus oocytes have been used extensively in studies of RNA nuclear transport. They are readily obtained from adult females, are large enough to be easily manipulated and microinjected, and the nucleus can be manually dissected from the oocyte (Fig. 1), allowing a simple determination of nuclear transport by measurement of the cytoplasmic and nuclear pools of the RNA under study.

Transport across the nuclear membrane occurs exclusively through aqueous channels formed by nuclear pore complexes (Görlich and Mattaj, 1996). Small molecules can freely diffuse through the pores, whereas larger molecules (≥ 40 kD) move through by an energy-dependent, facilitated process that requires specific molecular interactions (Görlich and Mattaj, 1996). Using microinjection into *Xenopus* oocytes as a functional assay, specific RNA structures and RNA-protein interactions have been implicated as requirements for facilitated nuclear transport of uridine-rich small nuclear RNAs (U snRNAs)

Methods and Tools in Biosciences and Medicine
Microinjection, ed. by J. C. Lacal et al.
© 1999 Birkhäuser Verlag Basel/Switzerland

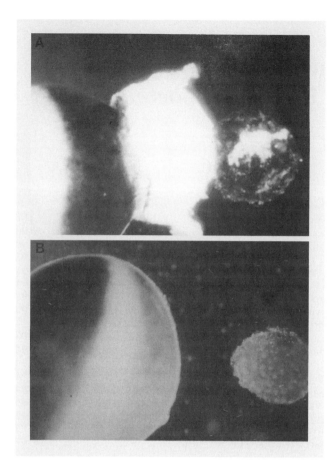

Figure 1 Manual dissection of the nucleus from a *Xenopus laevis* oocyte.

A. Nucleus with adhering cytoplasmic material and yolk (right) after extrusion from the pigmented animal hemisphere of the oocyte (left) B. Isolated nucleus (right) next to an intact oocyte (left, 1.2 mm diameter). (Photos by Chris Bain)

(Hamm and Mattaj, 1990; Michaud and Goldfarb, 1992; Terns et al., 1993; Fischer et al., 1994; Jarmolowski et al., 1994), mRNA (Dargemont and Kühn, 1992; Williams et al., 1994), tRNA (Tobian et al., 1985), signal recognition particle RNA (He et al., 1994), Y RNAs (Simons et al., 1996), ribosomal subunits (Bataillé et al., 1990; Pokrywka and Goldfarb, 1995), 5S RNA (Allison et al., 1993; Murdoch and Allison, 1996; Rudt and Pieler, 1996), and RNAs selected from combinatorial libraries (Grimm et al., 1997).

A number of nuclear localization sequence (NLS)-binding proteins have been characterized, along with other factors that comprise the pathway for nucleocytoplasmic transport (Görlich and Mattaj, 1996). Some of these and, as yet, unidentified factors appear to be specific for different classes of macromolecules. For example, most nuclear proteins, U6 snRNA, and 5S RNA enter the nucleus via a pathway that can be saturated by the SV-40 large T antigen NLS, whereas this signal does not interfere with nuclear import of U1-U5 snRNAs (Michaud and Goldfarb, 1992; Murdoch and Allison, 1996). Likewise, RNA export is mediated by class-specific, saturable factors (Jarmolowski et al., 1994;

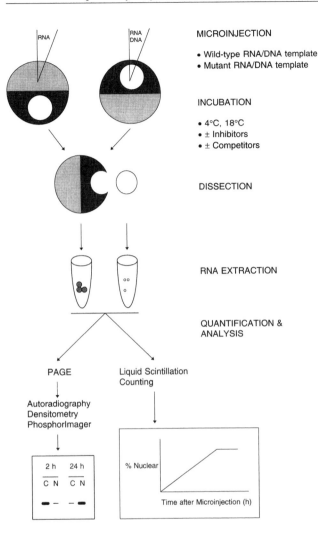

MICROINJECTION

• Wild-type RNA/DNA template
• Mutant RNA/DNA template

INCUBATION

• 4°C, 18°C
• ± Inhibitors
• ± Competitors

DISSECTION

RNA EXTRACTION

QUANTIFICATION &
ANALYSIS

PAGE

Liquid Scintillation
Counting

Autoradiography
Densitometry
PhosphorImager

2 h   24 h
C N   C N

% Nuclear

Time after Microinjection (h)

Figure 2 Flow diagram illustrating steps in the analysis of RNA nuclear transport in *Xenopus* oocytes after cytoplasmic or nuclear injection.

Pokrywka and Goldfarb, 1995; Izaurralde and Mattaj, 1995; Fridell et al., 1996; Simons et al., 1996).

In this chapter protocols are presented for:

1. Analysis of the kinetics of RNA nuclear transport, and the nucleotide sequence and structural requirements for transport.
2. Analysis of the mechanism of nuclear transport, as characterized by sensitivity to ATP depletion, chilling, wheat germ agglutinin (WGA) and anti-nucleoporin antibodies, and competition for shared transport factors (Fig. 2).

2   Materials

Solutions

- **Guanidine Thiocyanate Solution (Solution 1)**
    4 M guanidine thiocyanate (Boehringer Mannheim)
    25 mM sodium citrate, pH 7.0
    Dissolve at 65°C
    Store in a dark bottle at 4°C for ≤3 months
- **Phosphate-buffered Saline (PBS)**
    68 mM NaCl
    1.3 mM KCl
    4.0 mM Na_2HPO_4
    0.7 mM KH_2PO_4
    0.35 mM $CaCl_2$
    0.25 mM $MgCl_2$
- **RNA Extraction Solution (Solution 2)**
    Mix in a ratio of 1:1:0.1
    - TE-saturated phenol
    - Guanidine Thiocyanate Solution (**Solution 1**)
    - 2 M sodium acetate, pH 4.0
    Store in a dark bottle at 4°C for ≤3 months
    Just before use, add 720 µl of β-mercaptoethanol per 100 ml of Guanidine Thiocyanate Solution (**Solution 1**).
- **Nucleus Isolation Buffer (Solution 3)**
    25 mM Tris-HCl, pH 8.0
    10% glycerol
    5 mM $MgCl_2$
    2 mM dithiothreitol
- **Immunoprecipitation Buffer (Solution 4)**
    50 mM Tris-HCl, pH 7.4
    150 mM NaCl
    0.05% Nonidet P-40
    0.1 mM PMSF
    1 µg/ml aprotinin
    1 µg/ml leupeptin
    1 µg/ml pepstatin
    10 U/ml RNasin (Promega)

3   Methods

3.1   Kinetics of nuclear transport

To analyze the kinetics of nuclear transport, labelled RNA is injected either into the oocyte cytoplasm or nucleus, followed by measurement of radioactivity in each compartment over time. Alternatively, RNA export can be monitored after nuclear injection of DNA templates that are transcribed in the oocyte. For determination of nucleotide sequence and structural requirements for transport, transport kinetics of mutant RNAs are compared to wild-type RNA transport within the same batch of oocytes (e.g., Allison et al., 1993; Simons et al., 1996).

Protocol 1   Kinetics of nuclear transport

1.   Prepare oocytes for microinjection as described in Chapter 11.

2.   Microinject 1–10 fmol ^{32}P-labelled RNA (Allison et al., 1995) either into the cytoplasm (20–40 nl), or into the nucleus (10–20 nl). Alternatively, inject a DNA template (0.25 mg/ml) together with [α-^{32}P]GTP (0.1–1.0 μCi per oocyte), into the nucleus.

   Note:
   To monitor the accuracy of nuclear injections, mix RNA samples (1:1) with 20 mg/ml filter-sterilized blue dextran, a dye which is too large to exit the nucleus by diffusion and has no effect on nuclear transport or RNA-protein interactions. Ideally, also coinject an RNA with known localization as an internal control for the integrity of the nucleus, accuracy of microinjection, dissection, and sample recovery; e.g., U6ΔssRNA, an RNA that is retained in the nucleus (Simons et al., 1996).

3.   Incubate groups of microinjected oocytes at 18°C over a time-course from 0–24 h, to determine the rate and extent of transport. Always include a zero time point for nuclear injections as a control for potential leakage of samples during the injection process.

4.   At each time point, fix the oocytes at 4°C, a temperature at which facilitated transport does not occur. Select only oocytes that are healthy in appearance (uniform pigment, not mottled). Transfer oocytes from incubation medium (Chapter 11) to ice-cold 1% (w/v) trichloroacetic acid (TCA) in a sterile 35 × 10 mm plastic culture dish, using a "wide mouth" yellow tip (tip cut off to wider than the diameter of an oocyte). Use sterile watchmaker's forceps to open the oocyte with a single tear in the animal hemisphere, while applying pressure from the vegetal hemisphere with another pair of forceps. Roll out the nucleus and denude it of any adhering cytoplasmic material and yolk by gentle aspiration with an ultra-micro pipet tip (Fig. 1). For analysis of nuclear export, only collect nuclei and cytoplasm from successfully injected oocytes (those with blue nuclei).

5. Transfer enucleated oocyte "cytoplasms" and the corresponding nuclei to 1.5 ml tubes on ice.

6. Pool and homogenize 3–5 nuclei or cytoplasms in approximately 30 µl incubation medium, by vigorously pipetting with a yellow tip.

7. Add two uninjected "carrier" oocytes per sample of nuclei to aid in recovery of RNA.

8. Add 500 µl RNA Extraction Solution (**Solution 2**) and pipet to mix.

   Note:
   If oocyte fractions are added directly to RNA Extraction Solution they tend to harden rapidly and are difficult to homogenize.

9. Add 50 µl of chloroform/isoamyl alcohol (24:1) and vortex briefly to mix the phases. Keep the samples on ice for 15–30 min, then separate the phases by centrifugation at 10000 g for 20 min at 4°C.

10. Transfer the upper aqueous phase to a new 1.5 ml tube containing an equal volume of isopropanol. Mix gently by inversion and place at -20°C for ≥1 h.

11. Collect the RNA by centrifugation at 10000 g for 20 min at 4°C.

12. Wash the pelleted RNA once with 1 ml cold 70% ethanol, and collect by centrifugation at 10000 g for 10 min at 4°C.

13. Air dry the pellets for 10 min.

14. Resuspend the RNA pellet in 10 µl of 95% deionised formamide/0.05% xylene cyanol and bromophenol blue.

15. Heat at 95–100°C for 3 min to denature the RNA, place immediately on ice, and analyze by denaturing polyacrylamide gel electrophoresis (PAGE) (e.g., 8% polyacrylamide/8M urea gel in TBE buffer).

16. X-ray films of dried gels can be quantified by densitometry, or the dried gel can be quantified by PhosphorImager analysis.

17. Alternatively, resuspend the RNA pellets from Step 13 in 50 µl of 10 mM Tris-HCl, pH 7.6, 10 mM EDTA and add to 2 ml of biodegradable counting scintillant for liquid scintillation counting. Always visualize replicate samples by PAGE, to assess RNA stability.

There can be variability in synthetic activity between different batches of oocytes; thus, repeat experiments several times with oocytes from different animals, with at least three replicate samples per treatment and 3–5 oocytes per sample. Calculate the mean percentage of labelled RNA in the nucleus, plus or minus the standard error of the means. Graph kinetic data as "percent nuclear RNA versus time after microinjection." For analysis of mutant RNAs, calculate nuclear transport relative to the levels of wild-type RNA localized to the nucleus, within the same batch of oocytes.

3.2 Analysis of RNA-protein interactions

A complementary approach to the analysis of nuclear transport described above is to make use of RNA mutants that are defective in nuclear import or export, and to correlate this change in transport activity with a difference in protein binding between the mutant and wild-type RNAs (e.g., Allison et al., 1993, 1995; Rudt and Pieler, 1996; Izaurralde and Mattaj, 1995). When specific antibodies are available, RNA-protein interactions are analyzed by immunoprecipitation assay. Other methods include electrophoretic mobility shift assay (EMSA) (Allison et al., 1993, 1995) and UV-crosslinking (Grimm et al., 1997).

3.3 Preparation of oocyte extracts for immunoprecipitation assays

Protocol 2   Preparation of oocyte extracts for immunoprecipitation assays

1. 24 h after microinjection of the test substrate, isolate nuclei in **Solution 3**. Homogenize microinjected whole oocytes, cytoplasms, or nuclei in Immunoprecipitation Buffer (**Solution 4**).

2. Centrifuge whole oocytes and cytoplasms at 10000 g for 10 min at 4°C to pellet yolk and pigment, thereby decreasing non-specific binding during the immunoprecipitation assay.

3. Remove the supernatant to a fresh tube, mix gently, and aliquot 0.5 ml samples (20 oocyte equivalents) for analysis by a standard immunoprecipitation protocol (e.g., see Allison et al., 1993). Analyze RNA extracted from the immunoprecipitate and immunosupernatant fractions by denaturing PAGE.

3.4 Mechanism of nuclear transport

To determine whether RNA transport occurs via a facilitated process through the nuclear pore complex, transport assays are performed under conditions that allow passive diffusion but inhibit facilitated transport.

3.5 ATP depletion

Sensitivity to ATP depletion is one criterion for distinguishing facilitated transport from passive diffusion. Metabolic energy is required for the nuclear transport of U snRNAs (Michaud and Goldfarb, 1992), 5S RNA (Allison et al., 1993; Murdoch and Allison, 1996), ribosomal subunits (Bataillé et al., 1990), mRNA (Dargemont and Kühn, 1992), signal recognition particle RNA (He et al., 1994), and Y RNAs (Simons et al., 1996).

Protocol 3   ATP depletion

1. Oocyte ATP levels are rapidly depleted from 2 mM to 10 μM (Allison et al., 1993) by microinjection of the ATP-hydrolyzing enzyme apyrase (grade VIII, Sigma). The nuclear and cytoplasmic pools of ATP are not at equilibrium; thus, to deplete nuclear ATP and inhibit export, preinject 20 nl of 2.5 U/μl apyrase in phosphate-buffer saline (PBS) into the nucleus. Conversely, to deplete cytoplasmic ATP and inhibit import, preinject apyrase into the cytoplasm (Bataillé et al., 1990). As a control, inject PBS alone.

2. After 30 min incubation at 18°C, inject oocytes with the test substrate. After sufficient time to allow nuclear transport of the RNA being studied, quantify its subcellular distribution as described in Protocol 1.

   Note:
   After cytoplasmic injection of apyrase, the oocyte cytoplasm becomes quite fluid (Bataillé et al., 1990). Fixing in 1% TCA before dissection helps to avoid loss of cytoplasmic contents during enucleation. Prolonged treatment with apyrase can lead to high oocyte mortality.

Recent reports have demonstrated that the GTPase Ran plays a role in RNA transport in somatic cells (Izaurralde and Mattaj, 1995). Whether GTP hydrolysis is required during RNA transport in *Xenopus* oocytes may prove difficult to demonstrate directly. When GTP-γ-S, a non-hydrolyzable GTP analogue, was injected into oocytes at final intracellular concentrations of up to 80 μM, there was no apparent effect on 5S RNA import; however, at concentrations greater than 80 μM, the oocyte pigment became patchy and mottled after overnight incubation and the nucleus could not be dissected intact and, at concentrations of 400 μM to 4 mM (comparable to amounts used with somatic cells), the pigment was completely internalized within an hour, and the oocytes appeared as a yellowish ball which "cracked" when touched with forceps (L. A. Allison, unpublished observations).

3.6 Chilling

An additional criterion for facilitated transport is sensitivity to chilling, which slows enzyme-mediated reactions, but allows small molecules to diffuse freely across the nuclear envelope (Allison et al., 1993 and references therein).

Protocol 4   Chilling

1. Carry out injections and incubations at 4 or 18°C.

2. Incubate oocytes for the minimum amount of time for detectable transport to occur at 18°C. Prolonged treatment at 4°C results in high oocyte mortality.

3. After incubation, measure the radioactivity in the nucleus and cytoplasm as described in Protocol 1. As a control, incubate one group of oocytes at 4°C, followed by incubation at 18°C for an equal length of time.

Reversible inhibition shows that chilling blocks transport by inhibiting the activity of specific components of the transport machinery, rather than by preventing transport by way of nonspecific cellular damage.

3.7   Wheat germ agglutinin and anti-nucleoporin antibodies

Another criterion for distinguishing facilitated transport from passive diffusion is sensitivity to the lectin WGA (from *Triticum vulgaris*, Sigma). WGA binds specifically to N-acetylglucosamine residues on integral nuclear pore complex proteins (nucleoporins) and inhibits signal- and energy-dependent nuclear transport of RNA, although various classes of RNA show differential sensitivity to WGA (Michaud and Goldfarb, 1992; Allison et al., 1993; Grimm et al.,1997).

Protocol 5   Sensitivity to wheat germ agglutinin

1. Just before use, dissolve WGA in PBS at 2–20 mg/ml.

2. To allow time for interaction of WGA with nucleoporins, preinject WGA (20 nl) into the oocyte cytoplasm to inhibit import, and into the nucleus to inhibit export (Bataillé et al., 1990). Determine the optimal concentration by titration.

   Note:
   As a control for inhibitor specificity, preinject one group of oocytes with PBS alone and another group with a mixture of WGA and 500 mM N-acetylglucosamine (Sigma). The inhibitory effect of WGA should be negated in the latter, since N-acetylglucosamine competes with nucleoporins for binding to WGA. WGA is highly viscous and easily clogs injection needles. Check frequently to ensure that consistent size drops are being delivered.

3. After incubation for 1 h, inject oocytes with the test substrate, and analyze for nuclear transport as described above.

Much progress has been made recently in identifying components of the nuclear pore complex. Antibodies to specific nucleoporins differentially impair transport of multiple classes of RNA (Powers et al., 1997 and references therein). Concentrate affinity purified antibodies to 0.5–1.0 mg immunoglobulin (IgG)/ml. Inject 15–30 ng antibody, or control IgG, per oocyte together with the RNA or DNA template. If the antibodies contain significant ribonuclease activity, resulting in RNA degradation after injection, include RNasin (Promega) in the injectate (0.1 U per oocyte).

3.8   Competition for shared transport factors

Kinetically, a facilitated transport process is one whose rate is proportional to the concentration of the transported material, reaching a maximum when all carriers are saturated; in contrast, transport by simple diffusion is generally unsaturable. To establish whether the pathway for RNA nuclear transport possesses saturable binding components in common with other molecules, a comparison can be made between the RNA under study and molecules with differing nuclear transport requirements. By adding an excess of a competitor to the system, it is possible to test whether the transport process has a common component with the labelled RNA under study, since competition for a shared binding element will reduce the rate of the nuclear transport of the RNA relative to that in the absence of competitor.

1. Inject labelled RNA (1–10 fmol) alone or together with unlabelled competitor into the nucleus or cytoplasm. Commonly used competitors include:
   a) increasing amounts of unlabelled RNA (0.2 to 10 pmol) for self competition
   b) 5–10 pmol U snRNAs, tRNA, 5S RNA, or mRNA
   c) 1 mM dinucleotide analogue of the trimethylguanosine cap, m_7GpppG (Dargemont and Kühn, 1992)
   d) 15 µM p(lys) BSA (a synthetic nuclear protein which possesses the minimal SV-40 T antigen NLS; Michaud and Goldfarb, 1992), or histone H1 (Boehringer Mannheim)
   e) 0.25 µg synthetic peptide containing the nuclear export signal of HIV-1 Rev protein, a shared protein motif that may play an analogous role in mediating the nuclear export of both late HIV-1 RNAs and 5S RNA transcripts (Fridell et al., 1996; Simons et al., 1996).

2. After incubation, assay nuclear transport as described in Protocol 1, comparing transport kinetics in the absence of competitor and in the presence of competitor. Use saturability of transport by self-competition as a positive control for saturability of the transport process in each batch of oocytes.

4   Remarks

Several points should be kept in mind when interpreting transport data:

- The oocyte nucleus is about 20 times smaller by volume than the cytoplasm, or at least 10 times considering that about half the cytoplasmic volume is occupied by yolk platelets; thus, the amount of microinjected RNA present in the nucleus should be multiplied by a factor of 10–20 to appreciate the difference in terms of concentration between the nuclear and cytoplasmic compartments.
- Nucleocytoplasmic shuttling of the RNA of interest may occur. For example, nuclear transport of cytoplasmically microinjected 5S RNA is not a measure of nuclear import alone, since the 5S RNA is released from storage in the cytoplasm, migrates into the nucleus, is assembled into ribosomes, and then re-exported to the cytoplasm (Allison et al., 1993, 1995). Consequently, the saturability of 5S RNA import and export is not a simple function.
- The absence of transport of a mutant RNA could result not only from loss of interaction with a transport factor, but also from gain of a novel interaction with a cellular component that leads to its retention (Izaurralde and Mattaj, 1995; Grimm et al., 1997).
- A negative result in an immunoprecipitation assay does not necessarily mean the absence of an RNA-protein complex. Mutant RNA-protein complexes may have a different conformation not recognized by the antibody or the complexes formed may not be stable under the assay conditions. For example, 5S RNA mutants that were not immunoprecipitated with anti-TFIIIA antibodies were shown to have formed a stable complex with the protein by electrophoretic mobility shift assay (Allison et al., 1993).
- It may not always be possible to extrapolate findings from oocytes to other cell types; e.g., there are cell-specific differences in the signal requirements for U1 snRNA nuclear import in somatic cells and oocytes (Fischer et al., 1994).

References

Allison LA, North MT, Murdoch KJ et al. (1993) Structural requirements of 5S rRNA for nuclear transport, 7S ribonucleoprotein particle assembly, and 60S ribosomal subunit assembly in *Xenopus* oocytes. *Mol Cell Biol* **13**: 6819–6831.

Allison LA, North MT, Neville LA (1995) Differential binding of oocyte-type and somatic-type 5S rRNA to TFIIIA and ribosomal protein L5 in *Xenopus* oocytes: Specialization for storage versus mobilization. *Dev Biol* **168**: 284–295.

Bataillé NT, Helser T, Fried HM (1990) Cytoplasmic transport of ribosomal subunits microinjected into the *Xenopus laevis* oocyte nucleus: a generalized, facilitated process. *J Cell Biol* **111**: 1571–1582.

Dargemont C, Kühn LC (1992) Export of mRNA from microinjected nuclei of *Xenopus laevis* oocytes. *J Cell Biol* **118**: 1–9.

Fischer U, Heinrich J, van Zee K et al. (1994) Nuclear transport of U1 snRNP in somatic cells: differences in signal requirement compared with *Xenopus laevis* oocytes. *J Cell Biol* **125**: 971–980.

Fridell RA, Fischer U, Lührmann R et al. (1996) Amphibian transcription factor IIIA proteins contain a sequence element functionally equivalent to the nuclear export signal of human immunodeficiency virus type 1 Rev. *Proc Natl Acad Sci USA* **93**: 2936–2940.

Görlich D, Mattaj IW (1996) Nucleocytoplasmic transport. *Science* **271**: 1513–1518.

Grimm C, Lund E, Dahlberg JE (1997) *In vivo* selection of RNAs that localize in the nucleus. *EMBO J* **16**: 793–806.

Hamm J, Mattaj IW (1990) Monomethylated cap structures facilitate RNA export from the nucleus. *Cell* **63**: 109–118.

He X-P, Bataillé N, Fried HM (1994) Nuclear export of signal recognition particle RNA is a facilitated process that involves the *Alu* sequence domain. *J Cell Sci* **107**: 903–912.

Izaurralde E, Mattaj IW (1995) RNA export. *Cell* **81**: 153–159.

Jarmolowski A, Boelens WC, Izaurralde E et al. (1994) Nuclear export of different classes of RNA is mediated by specific factors. *J Cell Biol* **124**: 627–635.

Michaud N, Goldfarb DS (1992) Microinjected U snRNAs are imported to oocyte nuclei via the nuclear pore complex by three distinguishable targeting pathways. *J Cell Biol* **116**: 851–861.

Murdoch KJ, Allison LA (1996) A role for ribosomal protein L5 in the nuclear import of 5S rRNA in *Xenopus* oocytes. *Exp Cell Res* **227**: 332–343.

Pokrywka NJ, Goldfarb DS (1995) Nuclear export pathways of tRNA and 40S ribosomes include both common and specific intermediates. *J Biol Chem* **270**: 3619–3624.

Powers MA, Forbes DJ, Dahlberg JE et al. (1997) The vertebrate GLFG nucleoporin, Nup98, is an essential component of multiple RNA export pathways. *J Cell Biol* **136**: 241–250.

Rudt F, Pieler T (1996) Cytoplasmic retention and nuclear import of 5S ribosomal RNA containing RNPs. *EMBO J* **15**: 1383–1391.

Simons FHM, Rutjes SA, Van Venrooij WJ et al. (1996) The interactions with Ro60 and La differentially affect nuclear export of hY1 RNA. *RNA* **2**: 264–273.

Terns MP, Dahlberg JE, Lund E (1993) Multiple *cis*-acting signals for export of pre-U1 snRNA from the nucleus. *Genes Dev* **7**: 1898–1908.

Tobian JA, Drinkard L, Zasloff M (1985) tRNA nuclear transport: defining the critical regions of human tRNA$_i^{met}$ by point mutagenesis. *Cell* **43**: 415–422.

Williams AS, Ingledue TC III, Kay BK et al. (1994) Changes in the stem-loop at the 3′ terminus of histone mRNA affects its nucleocytoplasmic transport and cytoplasmic regulation. *Nucleic Acids Res* **22**: 4660–4666.

16 Functional Expression of G Protein-Coupled Receptors in *Xenopus laevis* Oocytes

P. de la Peña and F. Barros

Since the demonstration by Gurdon et al. (1971) that injection of foreign mRNA into *Xenopus laevis* oocytes leads to efficient and accurate translation of the encoded proteins, the oocyte system has been widely used as a tool for the study and/or cloning of receptors, ion channels and transporters. In fact, the oocyte is particularly useful as a general expression system because: i) its large size, which facilitates manipulations such as microinjection and electrode penetration, ii) the possibility to perform a variety of functional measurements using different technical approaches which include electrophysiological recordings, fluorescent dyes detection, binding of radioligands and flux measurements, and iii) the ability of the oocytes not only to express a wide variety of exogenous proteins, but also to perform postranslational processing of the newly synthesized entities (e.g. signal peptide cleavage, glycosylation, phosphorylation and subunit assembly). In this report we will focus mainly on the utility of the oocytes for expression of exogenous receptors. Some especial emphasis would be placed on studies from our laboratory centered in the expression of the thyrotropin-releasing hormone (TRH) receptor from rat adenohypophyseal cells.

Apart from the receptors that act as ligand-gated ion channels, a number of receptors for a variety of ligands have been expressed in *Xenopus* oocytes (Table 1). Although not exclusively limited to one class of receptors, the oocyte is particularly useful for the study of receptors that, like the TRH receptor, trigger the activation of phospholipase C (PLC). A scheme of the manipulation steps to follow with the oocytes and of the cellular cascade involved in the receptor-evoked responses is shown in Fig. 1. In this case, the expressed receptor will couple to an endogenous G protein, which in turn activates PLC, and this would promote an enhanced agonist-induced turnover of phosphatidylinositol 4,5-bisphosphate to inositol 1,4,5 trisphosphate (IP3) plus diacylglycerol. IP3 will bind to receptors on Ca^{2+}-sequestering intracellular organelles and evoke a release of Ca^{2+} into the cytosol. Finally, the increased cytosolic Ca^{2+} will lead to opening of Ca^{2+}-sensitive Cl^- channels in the plasma membrane. The resultant flux of Cl^- ions is the cause of a complex depolarization of the oocyte mem-

Methods and Tools in Biosciences and Medicine
Microinjection, ed. by J. C. Lacal et al.
© 1999 Birkhäuser Verlag Basel/Switzerland

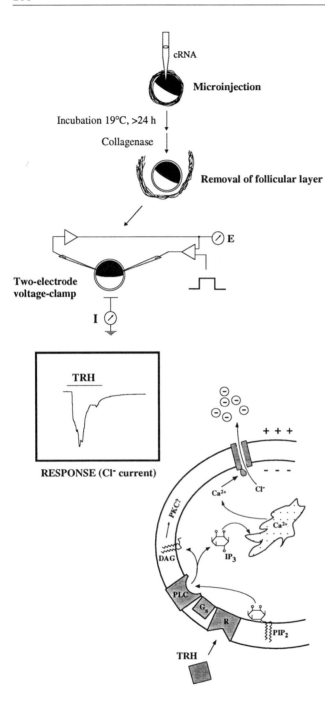

Figure 1 Diagram illustrating the principles for recording responses from oocytes expressing G protein-coupled receptors activating cellular phospholipase C and subsequently Ca^{2+}-activated Cl^- currents.

Microinjection of cRNA encoding the receptor is followed by incubation at 19°C for receptor functional expression. Removal of the oocyte follicular layer is performed mechanically with a pair of fine forceps following an optional treatment with collagenase (2 mg/ml collagenase IA for 60 min at room temperature). Whole-cell current measurements are performed at -60 to -80 mV with a two-microelectrode voltage-clamp at least 2 h after removal of follicular layer. Inward Cl^- current responses are recorded by placing the oocyte in a small groove of an experimental chamber (0.2 ml volume), and adding hormone (*TRH*) whereas the chamber is continuously perfused with saline at 3–5 ml/min (see de la Peña et al., 1992a,b for details). A scheme of the transduction cascade involved in elaboration of the oocyte response is schematically shown in the lower right part of the figure. See text for explanations.

Table 1   Receptors induced in *Xenopus* oocytes by microinjection of mRNA.

Receptor (ligand)	Reference
*IgC1 induction factor	Noma et al. (1986) *Nature* **319**: 640
#ACh $_{(m3)}$	Kubo et al. (1986) *Nature* **323**: 411
*Substance K	Masu et al. (1987) *Nature* **329**: 836
#α_2-adrenergic	Kobilka et al. (1987) *Science* **238**: 1406
#β_2-adrenergic	Kobilka et al. (1987) *J. Biol. Chem.* **262**: 15796
#β_1-adrenergic	Frielle et al. (1987) *Proc. Natl. Acad. Sci. USA* **84**: 7920
*5-HT$_{1C}$	Lübbert et al. (1987) *Proc. Natl. Acad. Sci. USA* **84**: 4322
Neurotensin	Hirono et al. (1987) *J. Physiol.* **382**: 523
Endothelin	Lory et al. (1989) *FEBS Lett.* **258**: 289
Angiotensin	Lory et al. (1989) *FEBS Lett.* **258**: 289
*Substance P	Yokota et al. (1989) *J. Biol. Chem.* **264**: 17649
*TRH	Straub et al. (1990) *Proc. Natl. Acad. Sci. USA* **87**: 9514
#fMet-Leu-Phe	Coats and Navarro (1990) *J. Biol. Chem.* **265**: 5964
*Bombesin	Spindel et al. (1990) *Mol. Endocrinol* **4**: 1956
#ACh $_{(m2)}$	Lechleiter et al. (1990) *EMBO J.* **9**: 4381
Somatostatin	White and Reisine (1990) *Proc. Natl. Acad. Sci. USA* **87**: 133
#Calcitonin	Lin et al. (1991) *Science* **254**: 1022
#Glutamate (metabotropic)	Masu et al. (1991) *Nature* **349**: 760
Vasopressin V2	Kimura (1991) *Biochem. Biophys. Res. Commun.* **174**: 149
Melatonin	Fraser et al. (1991) *Neurosci. Lett.* **124**: 242
*Platelet activating factor	Honda et al. (1991) *Nature* **349**: 342
#Thromboxane A$_2$	Hirata et al. (1991) *Nature* **349**: 617
Paratiroid hormone	Horiuchi et al. (1991) *J. Biol. Chem.* **266**: 4700
*Thrombin	Vu et al. (1991) *Cell* **64**: 1057
*Bradykinin B$_2$	McEachern et al. (1991) *Proc. Natl. Acad. Sci. USA* **88**: 7724
*5-HT$_3$	Maricq et al. (1991) *Science* **254**: 432
*Histamine H$_1$	Yamashita et al. (1991) *Proc. Natl. Acad. Sci. USA* **88**: 11515
#Choleocystokinin	Wank et al. (1992) *Proc. Natl. Acad. Sci. USA* **89**: 3125
*GnRH	Reinhart et al. (1992) *J. Biol. Chem.* **267**: 2128
Odorants	Dahmen et al. (1992) *J. Neurochem.* **58**: 1176
*Oxytocin	Kimura et al. (1992) *Nature* **356**: 526
*Vasopressin V1$_a$	Morel et al. (1992) *Nature* **356**: 523
*ATP P2	Lustig et al. (1993) *Proc. Natl. Acad. Sci. USA* **90** : 5113
*Ca^{2+}	Brown et al. (1993) *Nature* **366**: 575
#Growth hormone	Urbanek et al. (1993) *J. Biol. Chem.* **268**: 19025
Prostaglandin F2	Nakao et al. (1993) *J. Cell Physiol.* **155**: 257
#α_1-adrenergic	Blitzer et al. (1993) *J. Biol. Chem.* **268**: 7532
#Prostanoid IP	Rushmore et al. (1994) *J. Biol. Chem.* **269**: 12173
#Prostaglandin E2	Bastien et al. (1994) *J. Biol. Chem.* **269**: 11873
#Peptide C5a	Klos et al. (1994) *FEBS Lett.* **344**: 79
#Neuromedin B	Shapira et al (1994) *FEBS Lett.* **348**: 89
#GLP	Shapira et al (1994) *FEBS Lett.* **348**: 89

Receptor (ligand)	Reference
*Bradykinin B$_1$	Menke et al. (1994) *J. Biol. Chem.* **269**: 2158
#5-HT$_{2C}$	Chen et al. (1994) *Neurosci. Lett.* **179**: 100
VIC	Yoshimura et al. (1996) *Life Sci.* **58**: 171.
#Dopamine	Werner et al. (1996) *Mol. Pharmacol.* **49**: 656
#Opioid	Claude et al. (1996) *Proc. Natl. Acad. Sci. USA* **93**: 5715
Neuropeptide Y	Rimland et al. (1996) *Mol. Pharmacol.* **49**: 387

* Cloned by functional expression in oocytes. # Expressed in oocytes after cloning.
Ig: immunoglobulin; ACh: acetylcholine; 5-HT: serotonin; TRH: thyrotropin-releasing hormone; GnRH: gonadotropin releasing-hormone; GLP: gastrin-releasing peptide; VIC: vasoactive intestinal contractor.

brane in response to either foreign (Table 1) or endogenous (e.g., muscarinic, angiotensin II or sperm) receptors able to couple to this transduction cascade. However, from a practical point of view, recording of the Cl⁻ current under voltage-clamp is generally used as a sensitive and easily monitored measure of PLC activity and hence of receptor function (Fig. 1). The remarkable advantages of *Xenopus* oocytes for such electrophysiological recordings (reviewed in Snutch, 1988, Lester, 1988, Kushner et al., 1990 and Moriarty and Landau, 1990) and the enormous sensitivity of these measurements helped by the amplification contributed by the coupling cascade, constitutes a major reason for the extended use of the oocytes.

It is important to note that the reliability of the oocyte system for electrophysiological recordings makes it also an ideal candidate for cloning and studying functional properties and structure-function relationships of ionic channels. An extensive review on this topic, covering also the different methodologies used to maintain *Xenopus laevis*, to microinject mRNA, to remove oocyte covering layers, to express ionic channels and to perform two-electrode voltage-clamp and patch-clamp recordings in oocytes, can be encountered in Rudy and Iverson (1992). It is also interesting to indicate that the sensitive detection of ion currents due to operation of ion channels regulated by other second messengers different from Ca^{2+}, has been used for studies of receptors not coupled to cytoplasmic Ca^{2+} increases in oocytes. Thus, oocytes injected with mRNA for a cyclic-AMP regulated Cl⁻ channel (the cystic fibrosis transmembrane regulator (CFTR) protein), can be used for easy detection of expressed receptors that activate cyclic-AMP synthesis (Grygorczyk et al., 1995).

Although undoubtedly incomplete, the list showed in Table 1 includes receptors expressed in oocytes using mRNA either from native tissues or synthesized *in vitro* from previously cloned recombinant receptors. Thus, the oocyte system can yield very useful information about receptors on the following four specific issues:

• Receptor functional characteristics studied in a simple and well-controlled membrane environment.

- Cloning of functionally expressed receptors without any primary structure or purified protein availability.
- Validation of a cloned sequence as encoding a given receptor.
- Studies of structure-function relationships by expression of normal and mutationally altered receptors.

Injection of poly (A)⁺ mRNA from different tissues and cell lines allows for functional studies of receptors in a simple and well-controlled membrane environment

In this case, a direct characterization of the expressed protein structure cannot be made. However, since the coupling cascade of endogenous receptors to the adenylate cyclase and PLC has been identified and largely characterized in oocytes, the coupling characteristics to both of these transduction pathways after transplantation of foreign receptors can also be easily studied. This approach has been used to show that while liver [Arg[8]] vasopressin expressed receptors couple to PLC through a heterotrimeric G protein that is pertussis toxin (PTX)-sensitive, brain cholecystokinin-8 receptors stimulate the enzyme in a pertussis toxin-insensitive manner (Moriarty et al., 1989). The source of the transducer in these experiments is not known (i.e. endogenous to the oocyte *vs.* injected with the mRNA). However, an interesting feature of these results is that although the liver receptor utilizes a PTX-insensitive G-protein in its native environment, it utilizes a PTX-sensitive pathway in the oocyte. The nature of the transducer involved in responses to brain acetylcholine muscarinic and serotonin (5-HT) receptors has been also studied in rat brain mRNA injected oocytes (Kaneko et al., 1992). In this case, depletion of functional G-protein in the PLC pathway was achieved by injection of specific antisense oligonucleotides against $G_i\alpha$ or $G_o\alpha$. The results indicate that $G_o\alpha$ but not $G_i\alpha$, mediates brain 5-HT_{1c} receptor function and, in contrast, muscarinic receptors utilice $G_i\alpha$ rather than $G_o\alpha$ to activate PLC. Oocyte expression studies with mRNA from different tissues have been also used to explore the mechanism(s) of desensitization and/or down-regulation of 5-HT and TRH receptors (Oron et al., 1987, Mahlmann et al., 1989, Singer et al., 1990, Lipinsky et al., 1995), the different roles of IP3 and IP4 in the signal evoked by expression of TRH receptors (Mahlmann et al., 1989), or the influence of the expressed TRH receptor number on latency and amplitude of the hormone-induced response (Straub et al., 1989). Knockout of individual G proteins by microinjection of specific antisense oligonucleotides has been used also in our laboratory to demonstrate the functional coupling of the expressed TRH receptor to PLC *via* an oocyte endogenous G_s protein (de la Peña et al., 1995, see also Quick et al., 1994 and Stehno-Bittel et al., 1995). As a further refinement of using oocytes for studying the components of the coupling cascade to expressed receptors, a similar strategy has been employed recently, based in a subunit knock-out of individual G proteins by injection of specific antisense oligonucleotides or antibodies, in combination with elimination of free βγ dimers of G proteins by introducing βγ dimer-sequestrant agents in the oocyte (Stehno-Bittel et al., 1995). This led to the suggestion that βγ dimers are the predominant signaling molecule coupling receptors to activation of oocyte PLC. In this case, the role of the Gα subunit will be to determine the coupling specificity to receptors, but not to act as the coupling entity to the efector enzyme.

Expression of specific receptors after injection of mRNA from different sources allows cloning of the cDNAs or genes encoding receptor proteins for which primary structure information or purified protein are not available

As shown in Table 1, this approach has been very useful for isolation of several tens of receptors, but also of a panoply of ion channels and transporters. Use of a "classical" method for cloning a given protein relies on purification of the protein to generate antibodies to screen expression libraries, or to obtain a partial sequence in order to synthesize oligonucleotide probes to screen cDNA libraries. However, this approach would be only succesful if: i) the receptor or protein is present in abundant amounts in the source tissue, ii) a high affinity ligand susceptible to be radiolabelled is available to follow the purification process, and iii) the source tissue does not contain a number of similar receptors. When functional expression of a receptor after injection of oocytes is detected, an alternative cloning strategy is possible. In this case, the method uses the functional expression of the receptor for screening after injection of mRNA synthesized *in vitro* from fractions of a cDNA library increasingly enriched in the receptor cDNA (Masu et al., 1987). As a first step, the mRNA encoding the receptor is generally enriched by size fractionation. This enriched fraction is used to construct the cDNA library in which the proportion of clones encoding the desired receptor is increased. It could be useful to generate the library in a cloning vector which allows for insertion of big DNA fragments and also for *in vitro* synthesis of mRNA using the cDNA library as a template. After several cycles of fractionation of the library, cRNA synthesis using individual fractions, and detection of the receptor encoding clones in these fractions by functional expression, it is possible to obtain an individual clone encoding the receptor. As a variant of the method, a hybrid depletion procedure can be used (Lübbert et al., 1987, Jentsch et al., 1990). After constructing the cDNA library in a vector using the functionally active fraction of mRNA size-selected, single-stranded DNA derived from small groups of clones is used to deplete the mRNA corresponding to the desired receptor from total mRNA. The ability of the fractions to inhibit functional expression of the receptor is tested after injection of the mRNA "depleted" by every individual fraction into *Xenopus* oocytes. Again, this should eventually yield a single clone able to efficiently compete the expression of the active receptor. It is interesting to note that in case the active form of the receptor to be cloned is composed of more than one subunit, a point can be reached in which all subdivisions of an active fraction can be negative. In this case the subdivisions have to be recombined in various combinations until activity is regained. In fact, it should be kept in mind that since sampling is always based on functional expression, all selected clones must necessarily contain the functionally complete coding sequence.

An interesting possibility for cloning G protein-coupled receptors that use transduction elements not involved in elevations of intracellular Ca^{2+} or cyclic-AMP, has been recently opened by employing oocytes expressing an ion channel directly regulated by G proteins (Rimland et al., 1996). In this case, the electrophysiological detection of a channel operated by dissociated subunits of a receptor-activated G protein (either GTP-liganded α or free $\beta\gamma$ subunits), can be used to detect the expression of a receptor otherwise undetectable because its activation does not cause neither cyclic-AMP nor Ca^{2+} increases. Once the functional expression of the receptor that regulates the channel is achieved, successive cycles of fractionation and expression of the messages encoding the receptor (always coinjected with the mRNA encoding the channel), are used to obtain a single receptor clone as indicated above. Interestingly, choice of a channel regulated by a variety of G proteins (e.g. one activated by $\beta\gamma$ dimers common to different G proteins), opens wide possibilities to detect by this system a panoply of receptors interacting with different transducers.

Expression of DNAs cloned without any previous functional expression, can be used to validate identification of the cDNA as encoding the receptor (or protein) of interest

As an example, we started to clone the TRH receptor with an experimental design based on the succesive screening by functional expression in oocytes of a cDNA library generated from size-fractionated rat mammotropic GH_3 cells poly $(A)^+$. Before the screening was completed, the description of the primary structure of a cDNA encoding a TRH receptor from mouse thyrotropic cells allowed us to generate a probe by PCR, using two oligonucleotides of the mouse sequence and poly $(A)^+$ mRNA from GH_3 cells as a template. Use of this probe for screening of the cDNA library by the plaque hybridadization method yielded positive clones which apparently represented almost a full-length copy of the active TRH receptor mRNA. The cDNA clones were positively identified by functional expression in *Xenopus* oocytes as containing the message of the TRH receptor (de la Peña et al., 1992a) since: i) microinjection of sense RNA transcribed *in vitro* from the cDNA elicited electrophysiological responses to TRH identical to those originally observed with mRNA from GH_3 cells, ii) oocytes injected with antisense RNA transcribed *in vitro* from the cDNA did not show any response upon addition of TRH, iii) direct nuclear injection of the cDNA subcloned into a plasmid under control of the Herpes simplex virus thymidine kinase promoter elicited a response similar to that observed with TRH receptor mRNAs, iv) although the antisense RNA did not direct any functional expression in oocytes it was able to block the expression of TRH receptors when co-injected with poly $(A)^+$ mRNA from GH_3 cells, but not that of 5-HT receptors when co-injected with poly $(A)^+$ mRNA from rat brain, v) half-maximal TRH responses occurred at the same hormone concentration in oocytes injected with both GH_3 cells poly $(A)^+$ mRNA and *in vitro* transcribed RNA from the cDNA clone, and vi) the response of the oocytes injected with the *in vitro* transcripts of the cDNA clone was inhibited by chlordiazepoxide (a competitive

TRH receptor inhibitor in GH_3 cells) with similar potency as was the GH_3 cells poly (A)$^+$-directed functional expression.

Expression of mRNAs transcribed from normal as well as mutationally altered cloned cDNAs can be used to study structure-function relationships
In this case, both naturally occurring and site-directed mutagenesis generated receptor variants can be expressed to study the structural basis for each of the major functional attributes of the receptor: ligand binding, coupling to transducer or effector, and desensitization.

It has been shown that two metabotropic glutamate receptors generated by alternative splicing induce different patterns of Ca^{2+} release when expressed in oocytes (Pin et al., 1992). In contrast, we have observed that alternative spliced variants of the TRH receptor evoke undistinguishable responses upon expression (de la Peña et al., 1992b). Whether these similar responses are due to the lack of any difference in functional properties between the two TRH receptor isoforms, or to the inability of the oocyte system to reflect those differences in a measurable way, remains to be established.

Functional expression of mutationally altered receptors has been useful to localize both agonist and antagonist binding domains of the neurokinin-1 receptor (Fong et al., 1992). Similar studies with adrenergic receptors demonstrated a major role of the membrane-anchored core of the receptor on ligand binding (Kobilka et al., 1987), and localized important determinants of α_2- and β_2-adrenergic receptor agonist and antagonist binding specificity within the seventh membrane spanning domain (Kobilka et al., 1988). Recent work from our laboratory combining site-directed mutagenesis and functional expression in oocytes has identified an aminoacid residue in transmembrane domain III of the TRH receptor as an important determinant of hormone-receptor interactions. Different aminoacid substitutions at this position are able to decrease hormone-receptor interaction affinities by amounts ranging from 20 to more than one 100000-fold without affecting maximal response levels. A kinetic study of the expressed receptors allowed us to recognize an increased rate of dissociation of the hormone as the major cause of the affinity changes. Furthermore, the nature of the amino acid in this position seems to play a role, directly or indirectly, in conformational changes leading to receptor activation, and hence to signal transduction (del Camino et al., 1997).

Oocyte expression has been particularly useful to localize receptor regions involved in G protein-receptor coupling. It has been determined that the C-terminal portion of the third cytoplasmic loop and the N-terminal segment of the cytoplasmic tail appear to be critical for productive coupling of the α_2-adrenergic receptor to G proteins. In addition, two other areas of the receptor have been implicated in maintaining proper orientation of the G protein binding site (O'Dowd et al., 1988). The evaluation of the changes in kinetics and magnitude of oocyte responses after functional expression of muscarinic receptors normal, hybrid and deletion mutated, has been used to identify short sequences of the third cytoplasmic loop as the key regions involved in efficient and selective coupling to G proteins (Kubo et al., 1988, Lechleiter et al., 1990). In contrast, such determinants have been localized to the C-terminal end of the second intracellular loop and the segment located downstream of the seventh transmembrane domain in metabotropic glutamate receptors (Pin et al., 1994). Finally, it has been reported also that substitution of charged aminoacids or residues involved in amphipatic α-helical conformations disrupt coupling of M3 muscarinic receptor and G proteins (Duerson et al., 1993, Kunkel et al., 1993).

Possible constraints to use oocytes as an heterologous expression system for G protein-coupled receptors

Most of the conclusions to be reached after functional expression of receptors rely on the assumption that, since the oocyte contributes the biochemical cascade which couples the receptor to the response, any variation on response after expression of receptor variants would be related to variations in receptor structure. However, several constraints should be considered also regarding the use of the oocytes as expression system:

First, activation of oocyte responses by some expressed receptors is dependent on the presence of complementary factors which can be lost during the screening or the cloning procedure. This fact has been documented for formyl peptide and C5a receptors (Schultz et al., 1992a,b). Nevertheless, this has been also exploited in a positive way to identify the factor lacking in the oocyte by complementation of the cascade. Thus, the response evoked by these receptors is restablished by introduction of the α subunit of G_{16}, demonstrating also the specific coupling of the receptors to this specific G protein (Burg et al., 1995). An alternative situation is provided by detection of functional expression of certain ion channels in oocytes, in spite of lack of such an expression in other heterologous expression cell systems. This discrepancy allowed the identification of channel subunits necessary for normal activity of the channels, provided by the oocyte and that are not present in the other eukariotic cells used for expression. Subsequently, these subunits have been cloned from a *Xenopus* oocyte library (Hedin et al., 1996, Barhanin et al., 1996, Sanguinetti et al., 1996).

Second, high affinity, guanine nucleotide-sensitive agonist binding is undetectable after expression of $β_2$-adrenergic receptors in oocytes, probably due to presence of too high concentrations of receptors in relation to endogenous G protein concentrations (O'Dowd et al., 1988).

Third, it is known that the transduction pathway used by a given receptor depends on the specific properties not only of the receptor molecule, but also of the cell type (Vallar et al., 1990, Duzic et al., 1992). As stated before, PTX-sensitivity of transducer coupled to receptors can be changed after expression in oocytes (Moriarty et al., 1989). On the other hand, levels of expression can influence the signalling pathways used for a defined receptor (Milligan, 1993), and it has been argued that although receptors can be promiscuous in their choice of G protein, they appear to be so only when stimulated with non physiologically relevant concentrations and/or out of their native environment (Taylor, 1990).

Fourth, not only the expression, but some functional properties of the receptor itself can be lost or changed in the oocyte system. Thus, desensitization properties of serotonin 5-HT$_{1c}$ and muscarinic M1 receptors are modified by coinjection in oocytes of the receptor messages with low molecular weight fractions of rat brain mRNA (Walter et al., 1991). In summary, apart from their evident advantages as expression system, it is clear that these and other possible drawbacks should be also considered for interpretation of the functional expression results when oocytes are used.

Acknowledgements

The research cited from the authors laboratory was supported by grants PB90–0789 and PB93–1076 from C.I.C.Y.T. of Spain.

References

Barhanin J, Lesage F, Guillemare E et al. (1996) K$_v$LQT1 and IsK (minK) proteins associate to form the I$_{Ks}$ cardiac potassium current. *Nature* **384**: 78–80.

de la Peña P, Delgado LM, del Camino D, Barros F (1992a) Cloning and expression of the thyrotropin-releasing hormone receptor from GH$_3$ rat anterior pituitary cells. *Biochem J* **284**: 891–899.

de la Peña, P, Delgado, LM, del Camino, D, Barros, F (1992b) Two isoforms of the thyrotropin-releasing hormone receptor generated by alternative splicing have indistinguishable functional properties. *J Biol Chem* **267**: 25703–25708.

de la Peña P, del Camino D, Pardo LA et al. (1995) G$_s$ couples thyrotropin-releasing hormone receptors expressed in *Xenopus* oocytes to phospholipase C. *J Biol Chem* **270**: 3554–3559.

del Camino D, Barros F, Pardo LA, de la Peña P (1997) Altered ligand dissociation rates in thyrotropin-releasing hormone receptors mutated in glutamine 105 of transmembrane helix III. *Biochemistry* **36**: 3308–3318

Duerson K, Carroll R, Clapham. D. (1993) a-helical distorting substitutions disrupt coupling between m3 muscarinic receptor and G proteins. *FEBS Lett* **324**: 103–108.

Duzic E, Coupry I, Downing S, Lanier, SM (1992) Factors determining the specificity of signal transduction by guanine nucleotide-binding protein-coupled receptors. I. Coupling of a2-adrenergic receptor subtypes to distinct G-proteins. *J Biol Chem* **267**: 9844–9851.

Fong TM, Huang RC, Strader CD (1992) Localization of agonist and antagonist binding domains of the human neurokinin-1 receptor. *J Biol Chem* **267**: 25664–25667.

Grygorczyk R, Abramovitz M, Boie Y et al. (1995) Detection of adenylate cyclase-coupled receptors in *Xenopus* oocytes by coexpression with cystic fibrosis transmembrane conductance regulator. *Anal Biochem* **227**: 27–31.

Gurdon JB, Lane CD, Woodland HR, Marbaix G (1971) Use of frog eggs and oocytes for the study of messenger RNA and its translation in living cells. *Nature* **233**: 177–182.

Hedin KE, Lim NF, Clapham DE (1996) Cloning of a *Xenopus laevis* inwardly rectifying K+ channel subunit that permits GIRK1 expression of I_{KACh} currents in oocytes. *Neuron* 16: 423–429.

Jentsch TJ, Steinmeyer K, Schwarz G (1990) Primary structure of *Torpedo marmorata* chloride channel isolated by expression cloning in *Xenopus* oocytes. *Nature* 348: 510–514.

Kaneko S, Takahashi H, Satoh M (1992) Metabotropic responses to acetylcholine and serotonin of *Xenopus* oocytes injected with rat brain mRNA are transduced by different G proteins. *FEBS Lett* 299: 179–182.

Kobilka BK, MacGregor C, Daniel K et al. (1987) Functional activity and regulation of human β2-adrenergic receptors expressed in *Xenopus* oocytes. *J Biol Chem* 262: 15796–15802.

Kobilka BK, Kobilka TS, Daniel K et al. (1988) Chimeric α2,β2-adrenergic receptors: delineation of domains involved in effector coupling and ligand binding specificity. *Science* 240: 1310–1316.

Kubo T, Bujo H, Akiba I et al. (1988) Location of a region of the muscarinic acetylcholine receptor involved in selective effector coupling. *FEBS Lett* 241: 119–125.

Kunkel MT, Peralta EG (1993) Charged amino acids required for signal transduction by the m3 muscarinic acetylcholine receptor. *EMBO J* 12: 3809–3015.

Kushner L, Lerma J, Bennett MVL, Zukin RS (1989) Using the *Xenopus* oocyte system for expression and cloning of neuroreceptors and channels. *Meth Neurosci* 1: 3–28.

Lechleiter J, Hellmiss R, Duerson K et al. (1990) Distinct sequence elements control the specificity of G protein activation by muscarinic acetylcholine receptor subtypes. *EMBO J* 9: 4381–4390

Lester HA (1988) Heterologous expression of excitability proteins: route to more specific drugs?. *Science* 241:1057–1063.

Lipinsky D, Nussenzveig DR, Gershengorn MC, Oron Y (1995) Desensitization of the response to thyrotropin-releasing hormone in *Xenopus* oocytes is an amplified process that precedes calcium mobilization. *Pflügers Arch* 429: 419–425.

Lübbert H, Hoffman BJ, Snutch TP et al. (1987) cDNA cloning of a serotonin 5-HT1c receptor by electrophysiological assays of mRNA-injected *Xenopus* oocytes. *Proc Natl Acad Sci USA* 84: 4332–4336.

Mahlmann S, Meyerhof W, Schwarz JR (1989) Different roles of IP4 and IP3 in the signal pathway coupled to the TRH receptor in microinjected *Xenopus* oocytes. *FEBS Lett* 249: 108–112.

Masu Y, Nakayama K, Tamaki H et al. (1987) cDNA cloning of bovine substance-K receptor through oocyte expression system. *Nature* 329: 836–838.

Milligan G (1993) Mechanisms of multifunctional signalling by G protein-linked receptors. *Trends Pharmacol Sci* 14: 239–244.

Moriarty TM, Sealfon SC, Carty DJ et al. (1989) Coupling of exogenous receptors to phospholipase C in *Xenopus* oocytes through pertussis toxin-sensitive and -insensitive pathways. Crosstalk through heterotrimeric G-proteins. *J Biol Chem* 264: 13524–13530.

Moriarty TM, Landau EM (1990) *Xenopus* oocyte as model system to study receptor coupling to phospholipase C. In: *G proteins* (Iyengar R, Birnbaumer L, eds.) pp. 479–501, Academic Press, San Diego, CA, USA.

O'Dowd BF, Hnatowich M, Regan JW et al. (1988) Site-directed mutagenesis of the cytoplasmic domains of the human β2-adrenergic receptor. Localization of regions involved in G protein-receptor coupling. *J Biol Chem* 263: 15985–15992.

Oron Y, Straub RE, Traktman P, Gershengorn MC (1987) Decreased TRH receptor mRNA activity precedes homologous downregulation: assay in oocytes. *Science* 238: 1406–1408.

Pin J-P, Waeber C, Prezeau L et al. (1992) Alternative splicing generates metabotropic glutamate receptors inducing different patterns of calcium release in *Xenopus* oocytes. *Proc Natl Acad Sci USA* 89: 10331–10335.

Pin J-P, Joly C, Heinemann SF, Bockaert J (1994) Domains involved in the specificity of G protein activation in phospholipase C-coupled metabotropic glutamate receptors. *EMBO J* 13: 342–348.

Quick MW, Simon MI, Davidson N et al. (1994) Differential coupling of G protein α subunits to seven-helix receptors expressed in *Xenopus* oocytes. *J Biol Chem* 269: 30164–30172.

Rimland JM, Seward EP, Humbert Y et al. (1996) Coexpression with potassium channel subunits used to clone the Y2 receptor for neuropeptide Y. *Mol Pharmacol* **49**: 387–90.

Rudy B, Iverson LE (eds) (1992) Ion channels. *Meth Enzymol* vol. **207**.

Sanguinetti MC, Curran ME, Zou A et al. (1996) Coassembly of K_vLQT1 and minK (IsK) proteins to form cardiac I_{Ks} potassium channel. *Nature* **384**: 80–83.

Schultz P, Stannek P, Bischoff SC et al. (1992a) Functional reconstitution of a receptor-activated signal transduction pathway in *Xenopus laevis* oocytes using the cloned human C5a receptor. *Cell Signaling* **4**: 153–161.

Schultz P, Stannek P, Voigt M et al. (1992b) Complementation of formyl peptide receptor-mediated signal transduction in *Xenopus laevis* oocytes. *Biochem J* **284**: 207–212.

Singer D, Boton R, Moran O, Dascal N (1990) Short- and long-term desensitization of serotonergic response in *Xenopus* oocytes with brain RNA: roles for IP3 and protein quinase C. *Pflügers Arch* **416**: 7–16.

Snutch TP (1988) The use of *Xenopus* oocytes to probe synaptic communication. *Trends Neurosci* **11**: 250–256.

Stehno-Bittel L, Krapivinsky G, Krapivinsky L et al. (1995) The G protein $\beta\gamma$ subunit transduces the muscarinic receptor signal for Ca^{2+} release in *Xenopus* oocytes. *J Biol Chem* **270**: 30068–30074.

Straub RE, Oron Y, Gillo B et al. (1989) Receptor number determines latency and amplitude of the thyrotropin-releasing hormone response in *Xenopus* oocytes injected with pituitary RNA. *Mol Endocrinol* **3**: 907–914.

Taylor CW (1990) The role of G proteins in transmembrane signaling. *Biochem J* **272**: 1–13.

Vallar L, Muca C, Magni M et al. (1990) Differential coupling of dopamine D2 receptors expressed in different cell types. Stimulation of phosphatidylinositol 4,5-bisphosphate hydrolysis in LtK^- fibroblasts, hyperpolarization, and cytosolic-free Ca^{2+} concentration decrease in GH_4C_1 cells. *J Biol Chem* **265**: 10320–10326.

Walter AE, Hoger JH, Labarca C et al. (1991) Low molecular weight mRNA encodes a protein that controls serotonin $5-HT_{1c}$ and acetylcholine M1 receptor sensitivity in *Xenopus* oocytes. *J Gen Physiol* **98**: 399–417.

17 Expression of Glucose Transporters in *Xenopus laevis* Oocytes: Applications for the Study of Membrane Transporter Function and Regulation

M.J. Seatter and G.W. Gould

Contents

1 Introduction

The study of glucose transport and the kinetic properties of its transporters has received much attention in recent years, particularly upon the realisation that both facilitative and active glucose transport is mediated by families of homologous proteins. These proteins are the products of distinct genes, exhibit tissue-specific patterns of expression, and are endowed with distinct kinetic properties which serve to meet the particular need of different tissues for glucose uptake or release (Gould and Holman, 1993; Hediger and Rhoads, 1994). One major goal in this area has been to establish the precise kinetic properties of each specific transporter isoform. This is not a trivial exercise as, particu-

Methods and Tools in Biosciences and Medicine
Microinjection, ed. by J. C. Lacal et al.
© 1999 Birkhäuser Verlag Basel/Switzerland

larly for the facilitative (or GLUT) family, many tissues express more than one transporter isoform. This has complicated the analysis of whole cell transport assays, which are at the best of times difficult to perform.

One way to circumvent the problems inherent to somatic cell transport assays is to functionally express the transporter of interest in a heterologous host system and so study its properties in isolation. To this end, many such systems have been employed with varying degrees of success (Gould, 1994). Expression of members of the GLUT family has, for example, been reported in mammalian cell culture lines (Asano et al., 1992), bacteria (Thorens, 1988), Dictyostelium (Cohen et al., 1996), Sf9 insect cells (Cope et al., 1994) and *Xenopus* oocytes (Gould and Lienhard, 1988, Gould et al., 1991). Each of these systems offers the researcher different advantages and disadvantages. Cultured cell lines over-expressing transporters have been used with moderate success to study the functional properties of glucose transporters, but their chief drawback is the often high level of endogenous transporter expression which makes precise kinetic measurements difficult. Both bacteria and Sf9 cells have been shown to be capable of driving high levels of expression of GLUT1, however the functional state of the protein is compromised, such that performing kinetic analysis has proven fraught with difficulties.

In contrast, *Xenopus* oocytes offer an ideal system for the study of the properties of glucose transporters, and have been widely employed by scientists working on both facilitative glucose transporters (GLUTs) and sodium-dependent transporters (SGLTs) (Gould and Colville, 1994). The main advantages of oocytes in this regard include the low levels of endogenous glucose transport, the ease of injection and assay, and the opportunity to use a functional assay to rapidly screen mutants of transporters generated using molecular biology. These advantages are such that for most analyses of the functional properties of glucose transporters, oocytes are the system of choice. Below, we discuss some of the sorts of experiments which have been performed using oocytes to study glucose transport, and provide a detailed set of protocols employed in our laboratory for the study of glucose transport mediated by the GLUT family of facilitative glucose transporters.

2   Materials

Chemicals

- Collagenase II
- Unlabelled Sugar (e.g. 2-deoxy-D-glucose)
- Radiolabelled tracer sugar (e.g. [^{3}H] 2-deoxy-D-glucose, 1mCi/ml)
- Scintillation fluid.

Equipment

- Individual transporter cRNA-injected oocytes
- Individual dH$_2$0-injected oocytes

- Microfuge tubes
- 13.5 ml centrifuge tubes
- Scintillation vial inserts plus caps
- Aspirator

Solutions

- Barths buffer
- 1% SDS solution
- 1× PBS (150 mM phosphate buffered saline, pH 7.4) at 4°C.

3   Methods

3.1   Studies of the kinetic properties of facilitative glucose transporters

The most common sort of experiment one is likely to perform is, of course, an assay of the ability of each oocyte transporter population to accumulate one of its substrates. The simplest study to perform involves measuring the uptake of trace amounts of radioactively labelled sugar into oocytes which have been stored in Barths buffer, which contains no sugar, hence maintaining the cytosolic sugar concentration as low as possible. The external sugar concentration is varied for each measurement. The initial transport of substrate into the oocyte should be linear since the cytosolic sugar concentration is virtually zero. Only measurements of initial rates are recorded, before reverse-flow of sugar occurs. Therefore, what is being measured is the concentration at which the rate of uptake is half-maximal (K_m outside) which is related to the substrate binding affinity of the exofacial site, and the maximum velocity of sugar entry under infinitely high substrate concentrations (V_{max} entry). This type of assay is referred to as zero-*trans* uptake (Gould and Seatter, 1997).

The other most common type of assay is equilibrium exchange transport. Oocytes are preincubated and maintained with various concentrations of sugar and the uptake of tracer then measured. This method gives values of K_m and V_{max} for sugar exchange (Gould and Seatter, 1997).

It should be noted, however, that determination of V_{max} parameters is not useful unless the number of transporters at the oocyte plasma membrane can accurately be determined, giving the more useful k_{cat} value- the turnover number (Gould and Lienhard, 1988). To do this requires specific anti-transporter antisera, a sample standard (containing a known quantity of transporter protein), and a method for accurate purification of oocyte plasma membranes by subcellular fractionation. This latter condition is necessary to avoid inclusion of internally residing transporters (which are not involved in transport) in the quantitation. Unfortunately, achieving satisfactory results with this purification is technically difficult, due to the unique morphology of the oocyte, and not all laboratories have claimed to have achieved this goal.

Protocol 1   Zero-trans uptake assays

Detailed protocols for the injection of oocytes and preparation of cRNA are provided in the Chapters 1, 11, 15.

1.   Inject 50 nl of 1 mg/ml cRNA into the vegetal pole area of each oocyte. Inject an equal number of oocytes with dH_2O.

2.   Incubate oocytes for 48–72 h to allow expression and targeting of transporter proteins, replacing the bathing medium approximately every 12 h.

3.   Make up a fresh sugar substrate stock in Barths buffer at a concentration greater than 10 times that of the highest assay concentration required. For each assay condition, a 10× radiolabelled sugar stock is required. First decide on the total volume of radiolabelled sugar required, allowing 50 ml for each reaction (remembering to include dH_2O-injected controls) and an additional 50ml for determination of total counts added. In this laboratory, we typically perform assays in 500 µl volumes, thus necessitating 50 µl of isotope solution for each assay. Add concentrated substrate stock and trace radiolabelled substrate to microfuge tubes and dilute with Barths buffer to give a sugar concentration 10 times that of the desired reaction conditions with an activity of 10–20 mCi/ml. Each experiment would typically involve six different substrate concentrations, three which should be higher than the K_m value (if known) and three lower.

4.   Aliquot 10ml of each 10× radiolabelled sugar stock into five separate scintillation vials.

5.   If using hydrophobic inhibitors such as cytochalasin B, the oocytes should be incubated in a Barths buffer containing 1mg/ml type II collagenase for 30 min with agitation and then washed extensively in fresh Barths prior to assay. This removes the layer of follicular cells surrounding each oocyte, increasing the permeability of the oocyte to the inhibitor.

6.   Aliquot groups of 10 oocytes into 13.5 ml centrifuge tubes containing 0.45 ml Barths buffer at room temperature. Place these tubes in a rack adjacent to 10 scintillation vial inserts. If the effects of an inhibitor are to be measured, it should be added at this point, at 1.11× concentration, and a sufficient preincubation period allowed.

7.   Transport is initiated by the addition of 50 µl of 10× sugar solution to give a specified final sugar concentration with an activity of 1–2 µCi/ml. It is wise to stagger the time between initiation of transport for each group of oocytes by about 2 min to allow enough time for stopping each reaction. Mix by gentle agitation of the tube. Mix periodically throughout the incubation.

8. After a pre-designated time, usually 15 or 30 min, the transport of sugar into the oocytes is quenched by rapid aspiration of the media and addition of 5 ml ice-cold PBS. This is quickly repeated twice. If measuring transport by GLUTs the PBS solution should contain 0.1mM phloretin (a potent transport inhibitor which binds to the transporter, thus preventing efflux of substrate). Other inhibitors may be employed for other transporter species, such as phlorizin for SGLT1.

9. The oocytes should be dispensed individually into scintillation vials, using a pipetteman fitted with a yellow tip which has the end severed to a diameter slightly larger than the oocyte.

10. Add 1ml of 1% sodium dodecyl sulphate (SDS) to each vial.

11. The vials should be left for at least 1 h at room temperature and then vortexed to solubilise the oocytes. Note that Triton and other non-ionic detergents are not ideal for solubilising oocytes.

12. Add 4 ml of scintillation fluid and cap each vial, then shake well to mix the contents. Measure uptake by scintillation counting.

To determine the appropriate incubation time (step 8), simply perform an assay to measure the accumulation of a single, low concentration of radiolabeled sugar into transporter-injected and H_2O-injected oocytes at various time points. Deduct the cpm obtained for water-injected oocytes from those expressing the heterologous transporter. Plot a graph of uptake (cpm) vs. time (min). Choose a time point from this graph where substrate accumulation is linear with time and is easily accurately measurable.

For each group of transporter-injected oocytes assayed at a particular sugar concentration, a group of non-injected oocytes is assayed under identical conditions. This control gives a value for radioactive sugar transport by native oocyte transporters and simple diffusion plus background radiation levels. Hence, the average uptake in transporter-injected oocytes less the average transport in control (dH_2O-injected) oocytes gives a value (cpm/oocyte) corresponding to specific transport by heterologous transporters. Measurement of the mean cpm derived from 10ml of 10× substrate solution (step 4) can be used to calculate the quantity of radiolabelled substrate which produces 1 cpm (pmoles/cpm), allowing conversion of the mean quantity of transported sugar from cpm to pmoles/oocyte. A simple division of that figure by the transport period (step 8) gives an uptake rate with units expressed as pmoles/min/oocyte.

Transport with a variety of sugar concentrations are measured for each assay so that Lineweaver-Burk plots can be constructed. GLUT-mediated transport follows Michaelis-Menton kinetics such that:

$$1/v = K_m/(V_{max} \cdot S) + 1/V_{max}$$

Hence, by plotting 1/uptake rate (1/v) on the y-axis vs. 1/substrate concentration (1/S) on the x-axis, a straight line graph is obtained and the kinetics of transport can be derived.

When measuring the effects of inhibitors, both K_m and K_i values are determined. The plot is constructed as above to determine the K_m value, and a second plot constructed for transport under identical conditions, with inhibitor present. The K_i, which is equivalent to the inhibitor's dissociation constant, K_d, is the concentration of inhibitor which will inhibit 50% of transport, is measured using the same equation for both competitive and non-competitive inhibition which is:

$$\text{gradient of line (inhibitor)} = K_m / V_{max} \cdot (1 + [\,I\,] / K_i),$$

where [I] is the concentration of inhibitor used.

In general, most studies of GLUT activity employ 2-deoxy-D-glucose for such transport measurements. This analogue is metabolised inside the oocyte to 2-deoxy-D-glucose-6-phosphate by hexokinase (Colville et al., 1993a; Colville et al., 1993b; Gould and Lienhard, 1988; Gould et al., 1991). This phosphorylated species is not further metabolised, and thus accumulates inside the oocyte in a linear fashion with time, provided ATP is not rate limiting, as it is not a substrate for the GLUTs. Preliminary assays should involve a careful evaluation of this point (see above).

Protocol 2   Equilibrium exchange transport

The procedure is almost identical to that used for zero-*trans* uptake (**Protocol 1**) **except for the following differences.**

- Groups of 10 oocytes are preincubated at room temperature for 8–10 h in 0.45 ml of Barths buffer containing substrate at various concentrations.
- The reaction is initiated normally by addition of tracer to the external medium, and the rate of equilibration of the isotope determined.
- Measurements are made at 10, 20, 30 and 40 min for each substrate concentration. The equilibration is expected to be a first order process.
- The results are plotted as $-\ln[(C_m - C_t)/C_m]$ vs. time, where C_m is the radioactivity per oocyte after full equilibration of the oocyte water space and C_t is the radioactivity at time t.
- The rate constants (slopes) obtained from such plots (k_{obs}) are then used in Lineweaver-Burk plots to calculate the kinetic parameters, such as K_m. This is done by plotting $1/k_{obs} \cdot$ [substrate concentration] vs. 1/[substrate concentration], and proceeding as above.

Clearly for these types of study, a non-metabolised sugar analogue is employed. Most commonly, the analogue used is 3-O-methyl-D-glucose. For further details see Gould et al., 1991.

3.2 Studies of the kinetic properties of active glucose transporters

The procedure is similar to that outlined for zero-*trans* uptake by facilitative transporters (Hirayama et al., 1994; Pajor, 1992; PanayotovaHeiermann et al., 1995). SGLT1, for example, requires sodium in the external medium for active transport to occur. This is present in Barths buffer. The sodium-dependence of transport can be ascertained by full or partial replacement of the sodium chloride buffer constituent with 100mM choline chloride, to maintain the potential difference across the oocyte plasma membrane. Otherwise, transport measurements are performed as above. (For examples, see Hirayama et al., 1994; Pajor, 1992; PanayotovaHeiermann et al., 1995.)

3.3 Studies of the regulation of glucose transporters

Oocytes have been used by several laboratories to study the regulation of glucose transport. These studies fall into two classes, those which study the regulation of heterologous transporters, and those which use the endogenous oocyte transporter as a model system to study transporter regulation. We briefly discuss each of these below.

Regulation of heterologously expressed glucose transporters
There are two striking examples of studies which have addressed the regulation of transport activity using oocytes. Wright and colleagues have addressed the role of protein kinases in the control of SGLT1 function using oocytes (Hirch, 1996). In these studies, oocytes expressing the sodium-dependent glucose transporter, SGLT1, from a variety of species were expressed in oocytes by microinjection of *in vitro* transcribed cRNA. The functional properties of these expressed transporters were then studied after the activation of specific intracellular signalling pathways. Such approaches included the use of 8-bomoadenosine 3', 5'-cyclic monophosphate (a known activator of protein kinase A) and *sn*-1,2-dioctanoylglycerol (an activator of protein kinase C). Incubation of oocytes with these agents was shown to activate the appropriate kinase, and the subsequent effects of this activation on SGLT1 function could be readily assayed (Hirsch, 1996). Importantly, it should be remembered that oocytes, by virtue of their ease of microinjection, offer an ideal system to test the role of

any purified or recombinant protein. For example, it is possible to microinject activated protein kinases into oocytes (see below), and thus the effect of these proteins on the function of membrane transporters can be directly measured.

A further interesting use of oocytes to study glucose transporter regulation has come from the laboratory of Zorzano (Mora et al., 1995). It has long been established that insulin stimulates the movement ("translocation") of a pool of intracellular glucose transporters (GLUT4) to the plasma membrane of adipocytes and muscle (Holman and Cushman, 1994; James and Piper, 1994). The property of translocation is specific for the GLUT4 isoform, and is not exhibited to any significant degree by other GLUT isoforms. In an effort to address which features of GLUT4 dictate this response, and with a view to establishing a useful system to identify important components of the translocation machinery, Zorzano and colleagues (Mora et al., 1995) have shown that oocytes expressing GLUT4 exhibit large increases in IGF-I- and insulin-stimulated glucose transport, consistent with the translocation of this transporter to the surface of the oocyte. In contrast, oocytes expressing GLUT1 do not exhibit these large increases in transport rate. Subfractionation of the oocytes established that GLUT4 exhibited an insulin-dependent movement from an intracellular site to the oocyte plasma membrane. Such systems will be powerful tools in the search for the functionally important proteins which mediate insulin-stimulated glucose transport, and further illustrate the power of the oocyte as an experimental system.

Regulation of endogenous oocytes glucose transporters as a paradigm of transport regulation in somatic cells
Exposure of quiescent cells to growth factors causes activation of numerous signal transduction pathways which culminates several hours later in cell growth and division (Merrall et al., 1993a). Since growing cells have an increased energy requirement, one of the early events which follows re-entry into the cell cycle is the stimulation of glucose uptake, an effect which is common to all mitogens. In recent years there has been a tremendous increase in our effort to understand the molecular basis of growth factor action on glucose transport. The erythrocyte-type glucose transporter (GLUT1), which is expressed at detectable levels in most tissue types, all cell lines, transformed cells and tumour cells, is thought to play a "housekeeping" role in maintaining cellular levels of glucose transport (Merrall et al., 1993a). Recent studies have identified the native oocyte glucose transporter as a GLUT1 homologue, hence the oocyte offers a unique system for the study of the regulation of GLUT1 activity.

One of the main advantages of oocytes in this regard is that they are readily microinjected. We have used oocytes to study aspects of GLUT1 regulation, and shown that detectable chanto pharmacological intervention (Thomson et al., 1996).

Detailed review of these results is beyond the scope of this chapter, however, the key message is that oocytes do tolerate microinjection of proteins, peptides and drugs, the effects of which on glucose transport can be readily and simply assayed.

3.4 Studies on transporter structure

The oocyte system has also proved to be very useful for the characterisation of mutant transporters. Chimeras have been made between the closely related GLUT1/GLUT4 isoforms for the purpose of investigating targeting to different cellular compartments (Marshall et al., 1993). Transporters containing complementary sequences from GLUT1/GLUT4 (Dauterive et al., 1996, Due et al., 1995), GLUT3/GLUT4 (Burant and Bell, 1992), GLUT2/GLUT3 (Arbuckle et al., 1996) and GLUT2/GLUT4 (Buchs et al., 1995) have been constructed to investigate which protein sequences may be involved in dictating the unique substrate binding and selectivity of the different isoforms. This is feasible due to the high sequence identity shared between transporter isoforms. The functional effects of deleting variable lengths of C-terminal tail from GLUT1 and GLUT4 on transporter function and ligand binding have also been examined using oocytes as an expression vehicle to study the properties of the mutant species (Dauterive et al., 1996; Due et al., 1995).

Many conserved residues have also been targeted by site directed mutagenesis to investigate their function in substrate binding and/or transport, as well as ligand binding. The oocyte system has been utilised for the majority of these analyses. GLUT1 mutants which display an abnormal phenotype include Gln161Asn (Mueckler et al., 1994), Trp388Leu and Trp412Leu (Garcia et al., 1992). The reader is referred to Saravolac and Holman (1997) for detailed review of such studies.

Oocytes have also been used as a system for studies investigating the 2-dimensional membrane topology of transporters. This has been performed for both the active transporter SGLT1 (Turk et al., 1996) and GLUT1 (Hresko et al., 1994), using the technique of inserting glycosylation sequences into putative hydrophilic loops, and subsequently assessing the glycosylation status of the transporter, with the rationale that only loops exposed to the lumen of the endoplasmic reticulum (i.e. exofacially orientated) can be glycosylated. These types of study rely on the change in electrophoretic mobility on SDS-PAGE upon glycosylation. Hence, alterations in glycosylation pattern dependent upon the transmembrane orientation of the introduced glycosylation sequence reveal the relative transmembrane topology of the portion of the protein into which the site is introduced.

4   Remarks

What are the limitations of oocytes for the study of glucose transport?
Oocytes offer a powerful system for the determination of certain kinetic parameters of glucose transporters, a useful system for the study of their regulation, and have proven of use in aspects of structural determination. The relative ease with which large numbers of mutant transporters can be assayed
makes oocytes the system of choice for many studies of functionality. However,
there are some limitations which should be kept in mind. These include the dif-
ficulty in obtaining highly purified plasma membranes in a reproducible way
(which can, for example, limit determinations of turnover numbers (Arbuckle
et al., 1996)). Note however that some laboratories have described detailed
procedures which yield relatively pure plasma membranes (Marshall et al.,
1993). There are also some problems when oocytes are employed to examine
the effects of transport inhibitors, as the surrounding follicular cell layer can
interfere with such measurements. However, procedures for the removal of
the follicular cells does not preclude their use in this context.

Oocytes represent a powerful system with which to probe aspects of the
structure and function of membrane transporters. The arguments developed
above are equally applicable to a wide array of membrane transporters, and
such experimental approaches will no doubt rapidly enhance our understanding of membrane transport processes.

Acknowledgements

Research in GWGs laboratory is supported by The Wellcome Trust, The British
Diabetic Association, The Medical Research Council, Tenovus (Scotland), The
Agriculture and Food Research Council, and The Sir Jules Thorne Charitable
Trust. GWG is a Lister Institute of Preventive Medicine Research Fellow.

References

Arbuckle MI, Kane S, Porter LM, Seatter MJ,
   Gould GW (1996) Structure-function
   analysis of the liver-type (GLUT2) and
   brain-type (GLUT3) glucose transporters: expression of chimeric transporters
   in *Xenopus* oocytes suggests an important role for putative transmembrane
   helix VII in determining substrate selectivity. *Biochemistry* **35**: 16519–16527.

Asano T, Katagiri H, Takata K, Tsukuda K,
   Lin J-L, Ishihara H, Inukaio K, Hirano H,
   Yazaki Y, Oka Y (1992) Characterization
   of GLUT3 protein expressed in Chinese
   hamster ovary cells. *Biochem J* **288**:
   189–193.

Buchs A, Lu L, Morita H, Whitesell RR,
   Powers AC (1995) Two regions of GLUT2
   glucose transporter protein are responsible for its distinctive affinity for glucose.
   Endocrinology **136**: 4224–4230.

Burant CF, Bell GI (1992) Mammalian facilitative glucose transporters: evidence for similar substrate recognition sites in functionally monomeric proteins.*Biochemistry* 31: 10414–10420.

Cohen NR, Knecht DA, Lodish HF (1996) Functional expression of rat GLUT1 glucose transporter in *Dictyostelium discoideum Biochem J* 315: 971–975.

Colville CA, Seatter MJ, Gould GW (1993a) Analysis of the structural requirements for sugar binding to the liver, brain and insulin-responsive glucose transporters. *Biochem J* 194: 753–760.

Colville CA, Seatter MJ, Jess TJ, Gould GW, Thomas HM (1993b) Kinetic analysis of the liver-type (GLUT 2) and brain-type (GLUT 3) glucose transporters expressed in oocytes: substrate specificities and effects of transport inhibitors. *Biochem J* 290: 701–706.

Cope DL, Holman GD, Baldwin SA, Wolstenholme AJ (1994) Domain assembly of the GLUT1 glucose transporter. *Biochem J* 300: 291–294.

Dauterive R, Laroux S, Bunn SC, Chaisson A, Sanson T, Reed BC (1996) C-terminal mutations that alter the turnover number for 3-*O*-methylglucose transport by GLUT1 and GLUT4. *J Biol Chem* 271: 11414–11421.

Due AD, ZhiChao Q, Thomas JM, Buchs A, Powers AC, May JM (1995) Role of the C-terminal tail of the GLUT1 glucose transporter in its expression and function in *Xenopus laevis* oocytes. *Biochemistry* 34: 5462–5471.

Garcia JC, Strube M, Leingang K, Keller K, Mueckler MM (1992) Amino acid substitutions at tryptophan 388 and tryptophan 412 of the HepG2 (Glut1) glucose transporter inhibit transport activity and targeting to the plasma membrane in *Xenopus* oocytes. *J Biol Chem* 267: 7770–7776.

Gould G W (1994) *Membrane protein expression systems: a users guide*, Portland Press, London.

Gould GW, Colville CA (1994) *Expression of membrane transport proteins in Xenous oocytes.* Gould, GW Membrane protein expression systems: a users guide. Portland Press, London.

Gould GW, Holman GD (1993) The glucose transporter family: structure, function and tissue-specific expression. *Biochem J* 295: 329–341.

Gould GW, Jess TJ, Andrews GC, Herbst JJ, Plevin RJ, Gibbs EM (1994) Evidence for a role of phosphatidylinositol 3-kinase in the regulation of glucose transport in *Xenopus* oocytes. *J Biol Chem* 269: 26622–26625.

Gould GW, Lienhard GE (1988) Expression of a functional glucose transporter in *Xenopus* oocytes. *Biochemistry* 28: 9447–9452.

Gould GW, Seatter MJ (1997) In *Facilitative Glucose Transporters* (G.W.Gould, Ed.) pp 1–38, RG Landes, Austin.

Gould GW, Thomas HM, Jess TJ, Bell GI (1991) Expression of human glucose transporters in *Xenopus* oocytes: kinetic characterisation and substrate specificities of the erythrocyte, liver, and brain isoforms. *Biochemistry* 30: 5139–5145.

Hediger MA, Rhoads DB (1994) Molecular physiology of sodium-glucose cotransporters. *Physiol Rev* 74: 993–1026.

Hirayama BA, Loo DDF, Wright EM (1994) Protons drive sugar transport through the Na^+/glucose contransporter (SGLT1). *J Biol Chem* 269: 21407–21410.

Hirsch JR, Loo DDF, Wright EM (1996) Regulation of Na+/glucose cotransporter expression by protein kinases in *Xenopus* oocytes. *J Biol Chem* 271: 14740–14746.

Holman GD, Cushman SW (1994) Subcellular localisation and trafficking of the GLUT4 glucose transporter isoform in insulin-responsive cells. *BioEssays* 16: 753–759.

Hresko RC, Kruse M, Strube M, Mueckler M (1994) Topology of the GLUT1 glucose transporter deduced from glycoylation scanning mutagenesis. *J Biol Chem* 269: 20482–20488.

James DE, Piper RC (1994) Insulin resistance, diabetes, and the insulin-regulated trafficking of GLUT4. *J Cell Biol* 126: 1123–1126.

Marshall BA, Murata H, Hresko RC, Mueckler M (1993) Domains that confer intracellular sequestration of the GLUT4 glucose transporter in *Xenopus* oocytes. *J Biol Chem* 268: 26193–26199.

Merrall NM, Plevin RJ, Gould GW (1993a) Mitogens, growth factors, oncogenes and the regulation of glucose transport. *Cellular Signalling* 5: 667–675.

Merrall NW, Plevin RJ, Stokoe D, Cohen P, Nebreda AR, Gould GW (1993b) Mitogen-activated protein kinase (MAP kinase),

MAP kinase kinase and c-Mos stimulate glucose transport in *Xenopus* oocytes. *Biochem J* **295**, 351–355.

Mora S, Kaliman P, Chillaron J, Testar X, Palacin W, Zorzano A (1995) Insulin and insulin-like growth factor I (IGF-I) stimulate GLUT4 glucose transporter translocation in *Xenopus* oocytes. *Biochem J* **311**: 59–65.

Mueckler M, Weng W, Kruse M (1994) Glutamine 161 of GLUT1 glucose transporter is critical for transport activity and exofacial ligand binding. *J Biol Chem* **269**: 10533–10538.

Pajor AM, Wright EM (1992) Cloning and functional expression of a mammalian Na+/nucleoside cotransporter. *J Biol Chem* **267**: 3557–3560.

Panayotova-Heiermann M, Loo DDF, Wright EM (1995) Kinetics of steady-state currents and charge movements associated with the rat Na+/glucose cotransporter. *J Biol Chem* **270**: 27099–27105.

Saravolac EG, Holman GD (1997) In: *Facilitative glucose transporters* (Gould G W, Ed.) pp 39–61, RG Landes and Co, Georgetown, Texas.

Thomson FJ, Moyes C, Scott PH, Plevin R, Gould GW (1996) Lysophosphatidic acid stimulates glucose transport in *Xenopus* oocytes via a phosphatidylinositol 3'-kinase with distinct properties. *Biochem J* **316**: 161–166.

Thorens B, Sarkar HK, Kaback HR, Lodish HF (1988) Cloning and functional expression in bacteria of a novel glucose transporter present in liver, intestine, kidney and b-pancreatic islet cells. *Cell* **55**: 281–290.

Turk E, Kerner CJ, Lostao MP, Wright EM (1996) Membrane topology of the human Na+/glucose cotransporter SGLT1. *J Biol Chem* **271**: 1925–1934.

18 DNA Injection into *Xenopus laevis* Embryos as a Tool to Study Spatial Gene Activity

M. Kühl, M. Walter, J. Clement, H. Friedle, D. Wedlich and W. Knöchel

Content

1   Introduction

Pattern formation during vertebrate and invertebrate development becomes first apparent by the establishment of cell diversity. Different cells of the embryo can be distinguished by the distinct expression of different zygotic genes. In recent years, many developmentally important genes have been described in different organisms and, in many cases, orthologues of individual genes have been detected in a variety of species ranging from nematodes to humans. Especially in the South African clawed frog, *Xenopus laevis*, extensive work has been focused on genes being expressed in the marginal zone of the embryo, the prospective mesoderm. The marginal zone of the embryo is determined under the influence of maternally stored growth factors, e.g. of the TGF-β family, which are localized to the vegetal hemisphere. The dorsal-most region of the marginal zone represents the classical Spemann organizer. This part of the embryo will later give rise to the notochord, while more lateral domains will form muscle, heart and pronephros, and the ventral-most region will differentiate into blood cells. Mesoderm specific marker genes studied so far include dorsal specific genes like goosecoid (Niehrs et al., 1994), siamois

Methods and Tools in Biosciences and Medicine
Microinjection, ed. by J. C. Lacal et al.
© 1999 Birkhäuser Verlag Basel/Switzerland

(Lemaire et al., 1995), chordin (Sasai et al., 1994), noggin (Smith et al., 1992) or XFD-1/XFD-1' (pintallavis/XFKH1) (see Kaufmann and Knöchel, 1996). Ventral specific genes are the homeobox genes Xvent-1 and Xvent-2 (Gawantka et al., 1995; Onichtchouk et al., 1996), while Xbra (Smith et al., 1991) represents an early panmesodermal marker. Some of these genes are induced by activin even in the presence of cycloheximide (like goosecoid) and may thus represent direct targets of an activin A induced signaling cascade. Siamois, on the other hand, is only activated by different members of the Xvent family (Carnac et al., 1996). A challenge for the near future will be to elucidate the functional interaction of regulatory genes which are essential for pattern formation and to study the signal transduction cascades which are activated by extracellular growth factors.

One experimental strategy to look at the regulatory mechanisms underlying gene activation is the isolation of the promoter of an "interesting" gene and to study the spatial expression of reporter gene constructs in wild type or in experimentally manipulated embryos. Here, the *Xenopus* embryo provides an excellent experimental system which offers several advantages. Development of *Xenopus* proceeds extra-uterine and a large number of embryos at identical developmental stages are available after *in vitro* fertilization. Injected DNA constructs are used by the embryo when zygotic transcription starts after midblastula transition. If the promoter in question contains the control elements being necessary for correct spatial transcription, reporter gene activity is monitored according the pattern of wild type gene expression. No transgenic animal is needed to investigate gene expression from a reporter gene construct. Detection of reporter gene expression, either by whole mount *in situ* hybridization or by visualisation of β-gal activity, is simply achieved. In contrast to zebrafish embryos, dorsal-ventral polarity in *Xenopus* is already visible at four-cell stage embryos; this feature additionally allows to study spatial aspects of gene regulation. The influence of growth factors on the activity of a reporter gene construct can easily be studied in the animal cap assay. For this purpose, embryos are injected at the one- or two-cell stage into the animal half with the reporter gene construct. At blastula stage, the animal cap of the embryo is dissected and cultured. Exogenous growth factors can either be introduced by coinjection of their corresponding RNA into the cleavage stage embryo or just added as proteins to the culture medium. This experimental system is used as a routine system to study mesoderm induction in *Xenopus*, because animal cap cells which represent presumptive ectoderm will differentiate into mesodermal tissue upon addition of a mesoderm inducing factor.

Here we summarise the experimental procedures to investigate the spatial expression of genes and we present some data obtained with XFD-1' promoter/reporter gene constructs. Additionally, we present an example for tissue specific generation of antisense RNA and its consequences for embryonic development.

2   Materials

Chemicals

X-Gal	Gibco	15520–026
Paraformaldehyde	Sigma	P-6148
Glutaraldehyde	Serva	23115
Nonidet P-40	USB	19628
Na-deoxycholate	Sigma	D-6750

Solutions

- **PBS**
  137 mM NaCl
  2.7 mM KCl
  8.5 mM Na_2HPO_4
  1.5 mM KH_2PO_4, pH 7.3
- **MBS**
  88 mM NaCl
  10 mM KCl
  2.4 mM $NaHCO_3$
  0.8 mM $MgSO_4$
  0.33 mM $Ca(NO_3)_2$
  0.4 mM $CaCl_2$
  20 mM HEPES, pH 7.35
- **Fixative solution (Solution 1)**
  2% paraformaldehyde
  0.2% glutaraldehyde
  0.02% Nonidet p40
  0.1% Na-deoxycholate
  in 0.1× PBS
- **β-gal staining buffer (Solution 2)**
  5 mM $K_3Fe(CN)_6$
  5 mM $K_4Fe(CN)_6$
  2 mM $MgCl_2$, 0.01% Na-deoxycholate
  0.02% Nonidet P40
  1 mg/ml β-D-galactopyranoside (X-gal)

3   Methods

3.1   Design of reporter gene constructs

To study the spatial expression of a gene one first has to create suitable promoter/reporter gene constructs. We normally choose either LacZ, CAT or luciferase reporters by using the following vectors. phs3-lacZ contains the entire gene for LacZ under control of a mouse-derived heat shock promoter (Kothary et al., 1989). The heat shock element can be removed by Nco I/Sal I digestion and the promoter of choice can be inserted. The pEU-CAT vector (Piaggio and De Simone, 1990) has a very low activity by itself and was originally designed for studying very weak promoters. For luciferase as a reporter we use pGL3-basic (Promega).

The DNA used for injection experiments should be purified either by CsCl gradient centrifugation or by column chromatography (e.g. Qiagen-pure). *Do not* use simple miniprep DNA. The RNase added in the final step will lead to severe developmental defects. High concentrations of free nucleotides are also not favorable for the embryo. The concentration of the DNA in water is determined by UV spectroscopy. Either circular or linear DNA can be injected. We do normally not inject more than 10 nl in one cell of a four-cell stage embryo. Larger volumes may lead to malformations in bad batches of embryos which may result in false interpretations of reporter gene assays. Also, the amount of DNA injected per cell of a four-cell stage embryo may interfere with the quality of the embryo batch. In most cases 100 pg per blastomere represent the upper limit, very good batches of embryos also withstand 250 pg per blastomere. For CAT and luciferase reporters 20 pg are sufficient.

For injections we place the embryos in 4% Ficoll in 1× MBS and transfer them 1 h after the injection procedure into 0.1× MBS. Embryos are then cultured in 0.1× MBS at 18°C and staged according to Nieuwkoop and Faber (1975).

3.2   Tissue specific promoter activity detected by β-galactosidase as reporter

The gene product of the lacZ gene can be easily detected in whole embryos by a simple color reaction.

Protocol 1   Detection of lacZ gene product

1. Wash embryos three times prior to fixation with 1×PBS.

2. Fix embryos for 1 h on ice in fixative solution.

3. Wash three times in 1×PBS.

4. Place embryos in β-gal staining solution for 2 h to 24 h at 37°C.

5. After washing in 1× PBS fix embryos again for 1 h as described above and photograph.

6. Sections can be performed with a vibratome. We embed the embryos in 1.5% agarose and cut 50 μm sections.

This procedure should normally yield good results without any difficulties. You do not have to remove the vitelline membrane of the embryo before staining reaction. The added detergents are sufficient to ensure that the substrate will penetrate the whole embryo. A critical point for the method is the β-gal staining solution. The staining solution has always to be prepared freshly. In particular the cyanoferrat solutions may not be stored for more than a few weeks. We made the experience that the staining failed with old stock solutions. The X-gal is not brought into solution if you add it as the final component. However, this does not matter. Do not use X-gal that has been dissolved in DMF (as you will normally have from blue-white selections of bacteria). The use of a X-gal/DMF solution will result in a very high background. You will not be able to detect a specific promoter driven LacZ expression!

To demonstrate the power of this method we here present a β-gal staining of an embryo that has been injected with a XFD-1' promoter/LacZ reporter gene construct into both blastomeres of a two-cell stage *Xenopus* embryo. The -1102/+21 fragment of the XFD-1' promoter (Kaufmann et al., 1996) is sufficient to drive the LacZ expression within the endogenous expression domain of XFD-1' that is within the notochord and the neural floor plate (see figs. 1 A, B; Knöchel et al., 1992). Activation of the reporter is only observed within the dorsal part of the embryo and β-gal positive cells are confined to the dorsal midline (Fig. 1C). Cross-sections of these embryos also reveal the tissue specificity of the promoter fragment (see Figs 1D, E). The reporter gene is expressed in the notochord and the floor plate of neurula stage embryos. Therefore, the 1102 bp promoter fragment is sufficient to confer the correct spatial expression of the XFD-1' gene and presents a notochord/floor plate specific promoter.

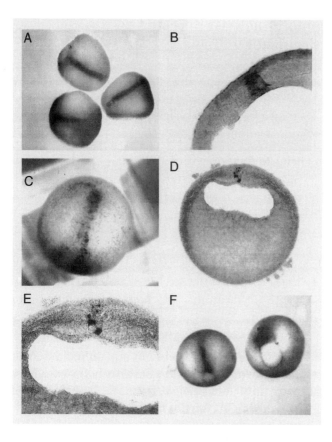

Figure 1 Spatial analysis of the XFD-1' promoter by reporter gene constructs.

A, B: whole mount *in situ* hybridization analysis for *Xenopus* XFD-1' transcripts within late gastrula/early neurula stage embryos (A). The transverse section in (B) shows expression in the notochord and the neural floor plate. C, D, E: β-gal activity in embryos which had previously been injected with the -1102 XFD-1' promoter/LacZ construct into both blastomeres at the two-cell stage. The transverse sections (D, E) clearly show that the promoter drives reporter gene expression to the notochord and the neural floor plate. F: detection of CAT reporter gene transcripts by whole mount hybridization in embryos which had been injected with the -1102 XFD-1' promoter/CAT construct at the two-cell stage. Note that the reporter gene is transcribed in the endogenous expression domain of XFD-1'.

Additionally, deletion mutant analysis of this promoter has revealed an activin response element and a BMP triggered inhibitory element, the latter preventing expression of this gene within the ventro/vegetal region of the embryo (Kaufmann et al., 1996). Note, that the expression of injected DNA is mosaic. This mosaic expression is a well known phenomenon after DNA injection into *Xenopus* embryos (Vize et al., 1991).

3.3 Analysis of spatial promoter activity by detection of reporter gene transcripts using whole mount *in situ* hybridization

CAT or luciferase reporter gene constructs can also be used in order to demonstrate a tissue-specific promoter activity. One simple procedure is to perform CAT enzyme assays (Gorman et al., 1982) in manually dissected parts of the embryo after injection of a promoter/CAT construct (Kaufmann et al., 1996). Using this method, and even at higher sensitivity by use of luciferase as reporter gene, different sections of embryonic material can be monitored for quantitative differences of promoter activities. In case of XFD-1', we have previously shown that the -1102/+21 fragment shows a dorsal-specific expression, while a -140/+21 fragment is also expressed in the ventral half of the embryo (Kaufmann et al., 1996). However, due to the sectioning procedure this method may not be very accurate and it does not allow expression studies on a single cell level. To overcome this problem we have analyzed the distribution of CAT reporter gene transcripts by whole mount *in situ* hybridizations (Harland, 1991). Embryos in Figure 1F were injected with the XFD-1' -1102 promoter/CAT reporter construct at the two-cell stage and the expression of the CAT gene monitored at late gastrula stage by whole mount *in situ* hybridization using a CAT antisense RNA as digoxygenin labelled probe. This probe was obtained by subcloning of a 257 bp Bam HI/Eco RI fragment of the pEU CAT vector (containing the 5' portion of the CAT gene) into the pSPT 18 vector to generate the antisense RNA. After the hybridization procedure, we additionally performed an incubation with RNase H. Note that the expression of the CAT gene strongly resembles the expression of the endogenous XFD-1' gene (Figs 1A, B) albeit its expression, as described above, is mosaic.

4   Remarks

Outlook: The use of tissue-specific promoters to inhibit gene function by means of antisense RNA

DNA injection experiments are not only restricted to studies of reporter gene activity to explore the spatial expression of genes and their interactions. We have investigated whether the tissue-specific expression of DNA constructs can also be used for functional gene studies. The restricted expression at a defined spatial domain of the embryo at a well defined time point might overcome some experimental disadvantages of the *Xenopus* embryo, e.g., the lack of gene knockouts. Most experiments done so far were performed with RNA injections affecting early embryonic development prior to the onset of zygotic transcription at midblastula transition. Injection of antisense RNA has been used to reduce goosecoid and BMP-4 mRNA at blastula stage (Steinbeisser et al., 1995). DNA injections in order to generate antisense RNA *in vivo* were mostly performed with ubiquitously activated strong promoters (Schmid et al., 1992; Nichols et al., 1995). By using this procedure the expression of DNA constructs was not restricted to the endogenous expression domain of the analyzed gene. The use of tissue-specific promoters might bypass this experimental hindrance. Such expression of DNA constructs under control of tissue specific promoters has to be seen in context with the attempts to create transgenic *Xenopus* by use of restriction enzyme mediated recombination (Kroll and Amaya, 1996).

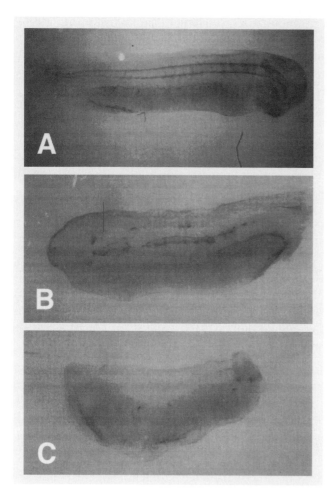

Figure 2 Expression of XFD-1 antisense RNA under control of the XFD-1' promoter.

One hundred pg of the described construct were injected in both dorsal blastomeres at the four cell stage. The notochord was stained using the monoclonal antibody Tor 70. Expression of XFD-1 antisense RNA within its endogenous expression domain results in loss of notochordal structures (B, C). The head structures are reduced or even missing. Control injection of an unrelated DNA construct revealed no abnormal phenotype (A).

Here, we show some initial studies to express antisense RNA of the XFD-1 gene under control of the XFD-1' promoter. The XFD-1/XFD-1' genes represent pseudoalleles that show identical expression profiles and their transcripts share 95 % identity at their 5' regions. We have cloned a 486 bp Eco RI/ Bgl II fragment of the XFD-1 cDNA clone in antisense direction downstream to the XFD-1' promoter. One hundred pg of this construct were injected into the dorsal blastomeres of four-cell stage *Xenopus* embryos. Embryos were analyzed for developmental defects at tadpole stage. Many embryos showed a ventralized and posteriorized phenotype (see Figs 2B, C). In most severe cases no head was formed. These defects were accompanied by abnormalities of the notochord as monitored by immunostaining with the notochord extracellular matrix antigen Tor 70. Embryos developed only patchy notochords (Fig. 2B), in severe cases the notochord was completely missing (Fig. 2C). These phenotypes were observed in 22.0 % of injected embryos (number of experiments = 3, number of embryos examined = 213). Control injections of equal amounts of an unrelated DNA (psh3-LacZ without the heat shock element) resulted only in

1.4 % of the observed phenotype (number of experiments = 3, number of embryos examined = 218). Total loss of the notochord was only observed in XFD-1 antisense injected embryos. This phenotype is reminiscent of the lethal phenotype described for HNF-3β knock-out mice. Loss of HNF-3β expression, the mammalian homolog to XFD-1, either leads to lack of node formation or results in embryos lacking notochord and head (Bally-Cuif and Boncinelli, 1997).

In conclusion, the use of tissue-specific promoters is a successful technique to study the function of zygotically expressed genes in *Xenopus* development. This method requires both tissue specificity and strong activity of the used promoter. Both have to be proven by β-gal or CAT expression before loss- or gain of function experiments are addressed.

References

Bally-Cuif L, Boncinelli E (1997) Transcription factors and head formation in vertebrates. BioEssays 19: 127–135.

Carnac G, Kodjabachian L, Gurdon JB, Lemaire P (1996) The homeobox gene siamois is a target of the Wnt dorsalsation pathway and triggers organiser activity in the absence of mesoderm. Development 122: 3055–3065.

Gawantka V, Delius H, Hirschfeld K, Blumenstock C, Niehrs C (1995) Antagonizing the Spemann organizer: role of the homeobox gene Xvent-1. EMBO J 14: 6268–6279.

Gorman CM., Moffat LF, Howard BH (1982) Recombinant genomes which express chloramphinicol acetyltransferase in mammalian cells. Mol Cell Biol 2: 1044–1051.

Harland RM (1991) *In situ* hybridization: An improved whole mount method for *Xenopus* embryos. Meth Cell Biol 36: 685–695.

Kaufmann E, Paul H, Friedle H, Metz A, Scheucher M, Clement J, Knöchel W (1996) Antagonistic sctions of activin A and BMP2/4 control dorsal lip-specific activation of the early response gene XFD-1' in *Xenopus laevis* embryos. EMBO J 15: 6739–6749.

Kaufmann E, Knöchel W (1996) Five years on the wings of fork head. Mech Dev 57: 3–20.

Knöchel S, Lef J, Clement J, Klocke B, Hille S, Köster M, Knöchel W (1992) Activin A induced expression of a fork head related gene in posterior chordamesoderm (notochord) of *Xenopus laevis* embryos. Mech Dev 38: 157–165.

Kothary R, Clapoff S, Darling S, Perry MD, Moran, LA, Rossant J (1989) Inducible expression of an hsp68-lacZ hybrid gene in transgenic mice. Development 105: 707–714.

Kroll K, Amaya E (1996) Transgenic *Xenopus* embryos from sperm nuclear transplantations reveal FGF signaling requirements during gastrulation. Development 122: 3173–3183.

Lemaire P, Garrett N, Gurdon JB (1995) Expression cloning of siamois, a *Xenopus* homeobox gene expressed in dorsal-vegetal cells of blastulae and able to induce complete secondary axis. Cell 81: 85–94.

Nichols A, Rungger-Brändle E, Muster L, Rungger D (1995) Inhibition of Xhox1A gene expression in *Xenopus* embryos by antisense RNA produced from an expression vector read by RNA polymerase III. Mech Dev 52: 37–49.

Niehrs C, Steinbeisser H, De Robertis E (1994) Mesodermal patterning by a gradient of the vertebrate homeobox gene goosecoid. Science 263: 817–820.

Nieuwkoop PD, Faber J (1975) Normal Table of *Xenopus* laevis (Daudin). 2nd edn. Elsevier/North-Holland Publishing Co., Amsterdam, The Netherlands.

Onichtchouk D, Gawantka V, Dosch R, Delius H, Hirschfeld K, Blumenstock C, Niehrs C (1996) The Xvent-2 homeobox gene is part of the BMP-4 signalling pathway controling dorsoventral patterning of *Xenopus* mesoderm. Development 122: 3045–3053.

Piaggio G, DeSimone V (1990) A new expression vector to study weak promoters. Focus 12: 85–86.

Sasai Y, Lu B, Steinbeisser H, Geissert D, Gont L, De Robertis E (1994) *Xenopus* chordin: A novel dorsalizing factor activated by organiser-specific homeobox genes. Cell 79: 779–790.

Schmid M, Steinbeisser H, Epperlein H-H, Trendelenburg M, Lipps HJ (1992) An expression vector inhibits gene expression in *Xenopus* embryos by antisense RNA. Roux́s Arch Dev Biol 201: 340–345.

Smith JC, Price BM, Green JBA, Weigel D, Herrmann BG (1991) Expression of a *Xenopus* homolog of Brachyury (T) is an intermediate-early response to mesoderm induction. Cell 67: 79–87.

Smith WC, Harland RM (1992) Expression cloning of noggin, a new dorsalizing factor localized to the Spemann organizer in *Xenopus* embryos Cell 70: 829–840.

Steinbeisser H, Fainsold A, Niehrs C, Sasai Y, DeRobertis E (1995) The role of gsc and BMP-4 in dorsal-ventral patterning of the marginal zone in *Xenopus*: a loss-of-function study using antisense RNA. EMBO J 14: 5230–5243.

Vize PD, Hemmati-Brivanlou A, Harland R, Melton DA (1991) Assays for gene function in developing *Xenopus* embryos. Meth Cell Biol 36: 368–388.

19 Transformation of Nematodes by Microinjection

S. Hashmi, G. Hashmi and R. Gaugler

Contents

1 Introduction

The free-living nematode, *Caenorhabditis elegans,* has been an important model system in the study of developmental and cell biology. Significant advances in mapping and sequencing the *C. elegans* genome have aided our understanding of fundamental biological processes. Development of an efficient method for gene transfer has been a key tool accelerating *C. elegans* research advances (Kimble et al., 1982; Stinchcomb et al., 1985., Fire, 1986; Mello *et al.*, 1991). Integrative transformation in *C. elegans* is reproducibly achieved after microinjecting DNA directly into maturing oocyte nuclei (Fire, 1986). Heritable extrachromosomal DNA transformation in *C. elegans* was first described by Stinchcomb et al. (1985) after microinjecting DNA into the gonad cytoplasm. DNA molecules injected into the cytoplasm of the *C. elegans* hermaphrodite gonad undergo a transient period of reactivity. This results in the formation of large heritable extrachromosomal structures that experience very little further rearrangement (Mello *et al.*, 1991). Germ cell nuclei in *C. elegans* develop initially in a syncytium (a multinucleate mass of protoplasm resulting from fusion of cells) and when cell membranes later envelop them, the exogenously added DNA is packaged into the oocyte. Transforming DNA is generally

Methods and Tools in Biosciences and Medicine
Microinjection, ed. by J. C. Lacal et al.

not integrated into the chromosomes, but rather is maintained as a concata-
mer (i.e. unite in a chain) of introduced sequences.

Microinjection methods for transferring foreign DNA into *C. elegans* have
helped advance basic studies of gene expression and have permitted the introduction of genes of interest for genetic studies (Fire, 1986; Stinchcomb et al.,
1985; Stringham et al., 1992) Mello et al. (1991) used a cloned mutant collagen
gene, *rol6 (su1006)* (Kramer et al., 1990) as a dominant genetic marker for
DNA transformation. They showed that large extrachromosomal array assembled directly from the injected molecules and that homologous recombination drives arrays` assembly. These workers suggested that the size of the as-
sembled transgenic structures determines whether or not they will be maintained extrachromosomally or lost. Low copy number extrachromosomal
transformation can be achieved by adjusting the concentration of DNA molecules in the injection mixture.

In addition to fundamental research using *C. elegans* as a model organism,
microinjection has contributed to genetic engineering of insect-parasitic nematodes intended for use as biological insecticides. Transgenic *Heterorhabditis
bacteriophora* have been generated carrying the *C. elegans hsp16*, *mec4*, and
hsp70 A genes (Gaugler and Hashmi, 1996). The transgenic *H. bacteriophora*
carrying a *C. elegans hsp70* A gene is highly thermotolerant compared with
the wild type, perhaps enhancing shelf-life.

In this chapter, we describe protocols for using microinjection in nematode
genetic transformation. The microinjection is performed in a clean isolated
area, preferably in a small room and involves several important steps:

I.   Culturing of nematode (Brenner, 1974; Hashmi and Gaugler, 1997)
II. Purification of plasmid DNA (Zhou et al., 1990)
III. Preparation of materials to be used in microinjection (Fire, 1986; Mello et
     al., 1991)
IV. DNA transformation (Fire, 1986; Mello et al., 1991; Hashmi et al., 1995a)

2   Materials

Chemicals

* Agarose
* Halocarbon oil (series 700, Halocarbon Products Corp., Augusta, SC, USA)

Equipment

* 100 µl micropipet (borosilicate capillary glass)
* 25 µl micropipet (borosilicate capillary glass)
* Micropipeter (0.5–10 µl and 10–100 µl sizes)
* 24 × 50 mm glass coverslip # 1 1/2 (VWR Scientific)

- Worm brush (to make a worm brush. Remove a single eyebrow hair and glue it to the tip of a needle using nail polish)
- Nitrogen tank.
- Stereomicroscope
- *Inverted microscopes:* The microinjection of DNA into nematodes is performed under a high power inverted microscope. Two popular microscopes are the Zeiss Axiovert 35 and the Nikon Diaphot-200. The microscopes are fitted with DIC optics and a circular sliding stage. A sliding stage with centered rotation is important for rapid positioning of nematodes for injection. Microscope should be placed in an area that is level and free of vibration.
- *Micromanipulators:* Used for very fine movements such as precisely injecting cells or microscopic organisms (e.g., nematode). Needle tips of 1 mm or less are used for nematode injection. Narishigi model MN-151 (Narishigi Co., Ltd. Japan) is a popular model and can be mounted directly to the stage of inverted microscopes.
- *Needle holder:* The injection needle is housed in an instrument holder attached to the micromanipulator.
- *Microinjector:* A microinjector system attaches the injection needle to a pressure source (N_2 gas). It includes a foot pedal controller and tubing that connects the needle assembly to the gas tank *via* a pressure gauge and a valve that can be manually opened and closed. DNA is forced through the needle under pressure from a tank of nitrogen gas.
- *Needle puller:* Utilized to make nematode injection needles. The popular Narishigi PN-3 (Narishigi Co., Ltd. Japan) is a horizontal puller controlled by a magnet (settings: heat 2.5, submagnet 82.9, main magnet 18.3). The appropriate settings for pulling needles must be determined by trial and error.
- Centrifuge

Solutions

- **1× TE buffer**
    10 mM Tris, pH 7.5
    1 mM EDTA
- **M9 buffer (Solution 1)**
    3 g KH_2PO_4
    6 g Na_2HPO_4
    5 g NaCl
    0.25 g $MgSO_4$
    per liter H_2O
    pH 7.2
    Autoclave and store at room temperature

3   Methods

3.1   Culturing of nematode

Protocol 1   Culturing of nematode

1. To start a nematode *in vitro* culture, collect thousands of nematodes in a 15–20 ml centrifuge tube.

2. Centrifuge at 2500–2800 rpm for 5 min and decant water.

3. Surface sterilize by adding 10–15 ml of 0.05% sodium hypochlorite solution in the tube, centrifuge, decant the supernatant (sodium hypochlorite).

4. Wash off sodium hypochlorite solution by adding 10–15 ml distilled autoclaved water in the tube.

5. Centrifuge and decant water; repeat this washing step twice.

6. After the final wash, remove all water from the tube and transfer concentrated nematodes (20 ml) onto a previously prepared agar plate (60 × 15 mm) pre-seeded with bacteria (*E. coli* OP50 for *C. elegans*; *Photorhabdus luminiscence* for *H. bacteriophora*) for nematode food and incubate at appropriate temperature (20 or 25°C depending on nematode species). Make several agar plates with nematodes.

7. After 3 days, collect adult hermaphrodites nematodes containing 4–6 eggs from agar plates by pouring approximately 10–15 ml M9 buffer (**Solution 1**) onto the plate (amount of M9 buffer may vary depending on the size of culture plate).

8. Shake the plate a little and transfer the nematode suspension to a sterile 50 ml centrifuge tube (size of tubes can vary).

9. To clean the nematodes of bacteria, centrifuge the tube at 2800–3000 rpm for 5–6 min.

10. Remove the supernatant and add 20–25 ml fresh M9 buffer, centrifuge again. Repeat this step 3–4 times.

11. Decant the supernatant from the tube, leaving just enough buffer in the tube so that nematodes do not desiccate. These nematodes are to be used for transformation. Buffer is used to reduce the osmotic stress.

3.2 Initial preparations

1. *Preparation of agarose pads:* The agarose pads are used to immobilize nematodes so they can be injected with considerable control without harm.
   - To make an agarose pad, prepare a solution of 2% agarose in water. The agarose solution should be kept molten (approximately 45°C) during preparation of pads.
   - Place 60 μl of molten 2% agarose onto a 24 × 50 mm glass coverslip.
   - Immediately place a second coverslip over the first, flattening the agarose drop.
   - When the agarose hardens, separate the two coverslips and save the coverslip holding the agarose pad.
   - Dry the pad at 80 °C for 15 min or leave over night at bench so that agarose become permanently affixed to the coverslip. Agarose pads can be stored for a few months at room temperature.

2. *Purification of plasmid DNA for injection:* There are several methods of DNA preparation for injection. A rapid minipreps is a preferred method (Zhou et al., 1990). Measure DNA concentration using the spectrophotometer, and dilute DNA in IX TE buffer. This DNA solution flows through injection needles effectively. The concentration of DNA used for microinjection can vary between 100–200 μg/ml (transformation frequency is 4× greater at plasmid concentration of 100 μl and above than at lower concentration). Store prepared DNA at –20°C for short-term storage or at -70°C for long-term storage). DNA used for transformation should contain a selectable marker for identification of transformants. *E. coli* β-galactosidase and green fluorescent protein (gfp) are commonly used reporter molecules for analyzing gene expression in nematodes. The histochemical techniques for analyzing organisms that contain a functional β-galactosidase fusion construct (Fire, 1992) or microscopic techniques for studying gfp fusion construct (Chalfie et al., 1994) are available.

3. *Preparation of DNA loading pipete:* Make a DNA loading pipet by heating a 100 μl micropipet in a Bunsen burner flame and pulling the ends apart as soon as the center melts. The "pulled out" region of the filling pipet must be thin enough to fit inside the injection needle and long enough to reach the needle tip. Or use a Hamilton Syringe to load DNA into injection needle.

4. *Preparation of injection needles:* Injection needles are made of a 25 μl micropipet (borosilicate capillary glass) using a needle puller. Since the needles tips obtained in this way are closed by the heat generated by the puller tips are broken open by gently touching agar debris. The best needles tip are not perfectly round: they have one edge that is fairly flat.

5. *Loading DNA into injection needle:* Dip the thin tip of the loading pipet into the DNA and allow capillary action to fill the pipette end. After a small amount is taken up, place the tip of the pipet into the open end of the injection needle and gently move it toward the needle tip until it touches the end

where the taper begins and can move no further. Blow gently into the pipet while slowly puling the pipet backward, forcing DNA into the tip of the needle. Load several injection needles prior to microinjection and keep them in a petridish. Discard all needles when finished with injection.

3.3 Materials assemblage before starting the injection

- Stereomicroscope
- Sterile forceps
- Sterile M9 buffer
- Agar plate pre-seeded with bacteria
- Worm brush
- Agarose pads and 25 µl micropipet for spreading oil on agarose pads
- Halocarbon oil
- Needle loaded with DNA
- Micropipeter (100 µl sizes)
- Nematodes to be injected

3.4 Microinjection procedure

Protocol 2   Microinjection procedure (nematodes)

1.   Attach the DNA-loaded needle into the collar of the instrument holder.

2.   Open the nitrogen tank valve to 30–40 p.s.i.

3.   Place a drop of halocarbon oil onto an agarose pad, spread the oil on pad by rolling capillary glass micropipet. Adding oil allows nematode adhesion to the pad.

4.   Viewing under the stereomicroscope, transfer 1–5 young adult hermaphrodite nematodes to a partially desiccated agar plate without bacteria to remove excess water, so that when they are transferred to an oil-covered agarose pad, they stick well to the pad.

5.   Using the worm brush, transfer nematode from the agar plate to an oil covered agarose pad, gently pressing nematodes to the pad.

6.   Transfer the agarose pad with nematodes to the inverted microscope, nematode side up.

7.   Under 100× magnification, bring the target nematode in focus.

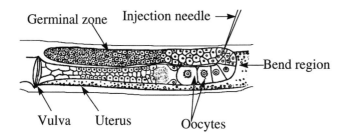

Germinal zone   Injection needle →

—Bend region

Vulva   Uterus   Oocytes

Figure 1 Schematic drawing of a nematode adult hermaphrodite gonad showing optimal placement of the microinjection needle.

8. Gently lower the needle until the tip comes into focus with the nematode. Switch to the 400× magnification and align the needle next to the gonad. The large gonad along the edge of the nematode should be in the same focal plane as the needle.

9. Insert the needle into the gonad either by using the micromanipulator or by moving the stage by hands. Press on the foot pedal to force DNA out of the needle and into the nematode gonad. The bend in the gonad should be the target of needle insertion (Fig. 1).

10. Move the nematode away from the needle and then release the pedal. Pressure release can pull the material into the end of the needle and clog it. The needle can be used for injection of many nematodes if it does not clog.

11. Bring the agarose pad with injected nematodes to the dissecting microscope and place a drop of M9 buffer on top of the nematode with a pipette. The nematode will float up into the buffer.

12. Using a worm brush, gently transfer the injected nematodes from the agarose pad onto a plate spread with bacteria. The nematodes dry out quickly during the injection process, therefore, steps 5–12 must be performed as quickly as possible.

13. Seal the plate with parafilm, and incubate at a temperature suitable for growth and reproduction.

14. After 24 h, transfer individual nematodes to a separate plate seeded with bacteria.

Examine progeny for the expression of the transformation marker (*E. coli* β-galactosidase or green fluorescent protein) to determine which nematodes carry introduced DNA in their germlines.

4   Troubleshooting

- *Low nematode survival after injection:* Reduce the time that the nematode is attached to the pad. Be careful in handling the nematodes, especially during transfer.
- *Poor DNA flow from the injection needle:* Determine whether the needle tip is open. Push on the foot pedal to observe the end of the needle to see if fluid comes out. See if the pressure gauge at the nitrogen tank shows proper pressure flow (30–40 p.s.i).
- *Low transformation efficiency:* Check DNA for purity by running on standard TAE Agarose gel or use the spectrophotometer. If needed, use selective precipitation to purify the DNA.

References

Brenner S (1974) The genetics of *Caenorhabditis elegans. Genetics* **77**: 71–94.

Chalfie M, Yuan, TU, Euskirchen, G et al. (1994) Green fluorescent protein as a marker for gene expression. *Science* **263**: 802–805.

Fire A (1986) Integrative transformation of *Caenorhabditis elegans. EMBO J* **5**: 2673–2680.

Fire A (1992) Histochemical techniques for locating *Escherichia coli* b-galactosidase activity in transgenic organisms. *Gata* **9**: 151–158.

Fire A, White-Harrison S, Dixon D (1990) A modular set of *LacZ* fusion for studying gene expression in *Caenorhabditis elegans. Gene* **93**: 189–198.

Gaugler R, Hashmi, S (1996) Genetic engineering of an insect parasite. In: *Genetic Engineering Principles and Methods* (JK Setlow, ed.), Vol. 18, PP. 135–155, New York: Plenum Press.

Hashmi S, Hashmi G, Gaugler R (1995a) Genetic transformation of an entomopathogenic nematode, *Heterorhabditis bacteriophora. J Invertebr Pathol* **66**: 293–296.

Hashmi S, Ling P, Hashmi G, Gaugler R et al. (1995b) Genetic transformation of nematodes using arrays of micromechanical piercing structures. *Biotechniques* **19**: 766–770.

Hashmi S, Gaugler R (1997) Injection of DNA into plant and animal tissues with microprobes. In: *Methods in Molecular Biology: Expression and Detection of Recombinant genes.* (R. Tuan, ed.) PP. 393–398, New Jersey: Humana Press

Kimble JJ, Hodgkin T, Smith et al. (1982) Suppression of an amber mutation by microinjection of a suppresser tRNA in *Caenorhabditis elegans. Nature* **299**: 456–458.

Kramer JM, French RP, Park EC et al. (1990) The *Ceanorhabditis elegans rol6* gene, which interacts with the sqt1 collagen gene to determine organismal morphology, encodes a collagen. *Mol Cell Biol* **10**: 2081–2089.

Mello CC, Kramer JM, Stinchcomb D et al. (1991) Efficient gene transfer in *C. elegans*: extrachromosomal maintenance and integration of transferring sequences. *EMBO J* **10**: 3959–3970.

Zhou C, Young A, Jong Y (1990) Miniprep in ten minutes. *Biotechniques* **8**: 173–175.

Guide to Solutions

Guide to Protocols

General Protocols

Microinjection Protocols

Assays

Troubleshooting Guide

Problems with oocytes

Problems with microinjection

Other problems

Index